AF352829

Extraction Strategies to Recover Bioactive Compounds, Incorporation into Food and Health Benefits

Extraction Strategies to Recover Bioactive Compounds, Incorporation into Food and Health Benefits

Special Issue Editors

María del Mar Contreras
Eulogio Castro

MDPI • Basel • Beijing • Wuhan • Barcelona • Belgrade • Manchester • Tokyo • Cluj • Tianjin

Special Issue Editors

María del Mar Contreras
University of Jaén
Spain

Eulogio Castro
University of Jaén
Spain

Editorial Office
MDPI
St. Alban-Anlage 66
4052 Basel, Switzerland

This is a reprint of articles from the Special Issue published online in the open access journal *Foods* (ISSN 2304-8158) (available at: hhttps://www.mdpi.com/journal/foods/special_issues/extraction).

For citation purposes, cite each article independently as indicated on the article page online and as indicated below:

LastName, A.A.; LastName, B.B.; LastName, C.C. Article Title. *Journal Name* **Year**, *Article Number*, Page Range.

ISBN 978-3-03928-969-1 (Pbk)
ISBN 978-3-03928-970-7 (PDF)

Contents

About the Special Issue Editors

María del Mar Contreras, Ph.D., holds a Ph.D. in Food Science and Technology and Chemical Engineering from the Autonomous University of Madrid, Spain. Currently, she is a postdoctoral researcher at the University of Jaén (Spain). She has worked in different Spanish institutions, including the Institute of Food Science Research (CIAL), University of Granada and University of Cordoba. She has been involved in the development of innovative approaches for characterizing bioactive components from foods and agro-industrial byproducts, for tackling issues such as obtaining novel functional ingredients, authentication, and bioavailability. Her current interest surrounds the valorization of agricultural and agro-industrial byproducts through obtaining bioactive compounds and their characterization, as well its integration in biorefinery schemes.

Eulogio Castro, Ph.D., holds a Ph.D. in Chemical Engineering from the University of Granada, Spain. He is currently a Full Professor of Chemical Engineering at the University of Jaén, Spain. He was also Invited Researcher at the École Nationale Supérieure de Chimie de Toulouse, France, and Visiting Assistant Professor at the University of Florida, USA. His main research interests are focused on the conversion of biomass into added-value products such as biofuels, renewable chemicals or biomaterials and also the techno-economic and environmental issues related to the development of the biorefinery concept.

Preface to "Extraction Strategies to Recover Bioactive Compounds, Incorporation into Food and Health Benefits"

Over the last years, numerous studies have addressed bioactive compounds from foods. Agricultural and agro-industrial byproducts have a negative environmental impact in several regions worldwide, since solutions for their valorization are either not or poorly implemented. Notably, these by-products are rich sources of valuable bioactive compounds, whose recovery could serve to obtain new natural additives, nutraceuticals, and functional ingredients for pharmaceutical, cosmetics, and food industries. To bring together new knowledge on bioactive compounds, especially from agri-food byproducts, when applied to the food industry, we edited this Special Issue entitled "Extraction Strategies to Recover Bioactive Compounds, Incorporation into Food, and Health Benefits" (URL: https://www.mdpi.com/journal/foods/special_issues/extraction).

The first objective of this Special Issue was to address new methods to sustainably recover bioactive compounds, not only from food matrices but also from agri-food byproducts. This valorization method could be complementary to biorefinery to also obtain bioenergy, biofuels, and/or other biocompounds, which merits investigation to migrate toward a circular bioeconomy model. In this line, some of the studies presented here are:

- Distribution of Free and Bound Phenolic Compounds in Buckwheat Milling Fractions

- Phenolic Compounds from Sesame Cake and Antioxidant Activity: A New Insight for Agri-Food Residues' Significance for Sustainable Development

- Integrated Process for Sequential Extraction of Bioactive Phenolic Compounds and Proteins from Mill and Field Olive Leaves and Effects on the Lignocellulosic Profile

The second objective of this Special Issue was to present studies that support the presence of the bioactive compounds in the food matrix and that provide health benefits. In this topic, the following studies were published:

- Biocompounds Content Prediction in Ecuadorian Fruits Using a Mathematical Model

- Bioprocessed Production of Resveratrol-Enriched Rice Wine: Simultaneous Rice Wine Fermentation, Extraction, and Transformation of Piceid to Resveratrol from Polygonum cuspidatum Roots

We would like to thank the professional staff at MDPI for their efforts that made this Special Issue and this edition possible. We are at your disposal for any further information on this edition.

María del Mar Contreras, Eulogio Castro
Special Issue Editors

Editorial

Extraction Strategies to Recover Bioactive Compounds, Incorporation into Food and Health Benefits: Current Works and Future Challenges

María del Mar Contreras [1,2,*] **and Eulogio Castro** [1,2,*]

[1] Department of Chemical, Environmental and Materials Engineering, Universidad de Jaén, Campus Las Lagunillas, 23071 Jaén, Spain

[2] Centre for Advanced Studies in Earth Sciences, Energy and Environment (CEACTEMA), Universidad de Jaén, Campus Las Lagunillas, 23071 Jaén, Spain

* Correspondence: mcgamez@ujaen.es or mar.contreras.gamez@gmail.com or mmcontreras@ugr.es (M.d.M.C.); ecastro@ujaen.es (E.C.)

Received: 25 March 2020; Accepted: 26 March 2020; Published: 30 March 2020

There are numerous studies in the literature about bioactive products (extracts, essential oils, oleoresins, hydrolysates, etc.), exhibiting numerous health benefits. Foods of plant and animal origin [1,2], medicinal plants [1,3], and marine products [4] can be their sources. We recognize the need to look for more efficient agroindustry because of unsustainable current practices, the population factor, and the state of natural resources. This requires us to move towards a more sustainable circular model that uses renewable resources, i.e., bioeconomy. Therefore, this implies a shifting to produce more food and to valorize agricultural byproducts and all agro-industrial derived streams through obtaining bio-based chemicals and products (including bioactive products), as well as bioenergy and/or biofuels [5]. To bring together new data to expand the information on bioactive compounds when applied to the food industry, we edited this special issue on "Extraction Strategies to Recover Bioactive Compounds, Incorporation into Food, and Health Benefits" (URL: https://www.mdpi.com/journal/foods/special_issues/extraction).

One of the objectives of the special issue was to address new ways to extract bioactive compounds from raw materials, including agri-food waste, sustainably, and mainly using food-grade conditions. In this context, four articles have been accepted, applying several extraction strategies, mostly assisted by ultrasound, to recover and ascertain bioactive compounds in Ecuadorian fruits [6] and buckwheat flour [7], as well as in agro-industrial byproducts from the processing of pseudo-cereal food [7], sesame [8], and the olive oil industry [9]. In the first work, Martín-García et al. (2019) determined the free and bound phenolic compounds in dehulled whole buckwheat flour and three milling fractions: light flour, bran flour, and middling flour. For that, an ethanol-water solution was used to extract free phenolic compounds, being assisted by ultrasound, and diethyl ether/ethyl acetate was applied to recover bound phenolic compounds after an alkaline treatment of the residual fraction. The most abundant free phenolic compounds were rutin and epiafzelchin–epicatechin-*O*-dimethylgallate, whereas the most abundant bound phenolic compounds were catechin and epicatechin in all buckwheat flours. The highest content of bound phenolic compounds (around 0.7 g/kg) was found in middling and bran flours, highlighting that both entire byproducts fractions could be used to develop functional foods owing to the beneficial health benefits of buckwheat phenolic compounds. The work by Mekky et al. [8] gives new insight into the phenolic composition of sesame and in particular of the residual cake after oil extraction using reversed-phase high-performance liquid chromatography coupled to diode array detection and quadrupole-time-of-flight-mass spectrometry. The characterized compounds belonged to several classes, namely, hydroxybenzoic acids, hydroxycinnamic acids, flavonoids, and lignans. Their findings suggest that the antioxidant activity of the sesame cake extract was not only promoted by sesamol and lignans, which are the phenolic compounds previously reported in sesame, but *C*-glycosides and other compounds could also contribute to this bioactivity. Furthermore, Contreras and coworkers [9] have proposed an integrated scheme to recover oleuropein and other olive leaves antioxidants in the first extraction step assisted by ultrasound, proteins in a second extraction step along with mannitol and arabinose oligomers, and finally, a cellulose-enriched fraction was obtained residually, which can be subject to subsequent valorization (e.g., for obtaining biofuel).

The extracts and fractions containing bioactive compounds can be addressed to formulate nutraceuticals, preservatives, such as antioxidant and antimicrobial additives, or functional ingredients for functional foods. Moreover, some plant foods can contain a high content of bioactive components, which can be predicted using mathematical models and based on color measurements, as showed Llerena and coworkers [6]. Yang et al. [10] directly obtained a resveratrol-enriched rice wine with enhanced antioxidant activity by adding *Polygonum cuspidatum* root powder, as a source of resveratrol, and fermentation. After 10 days of co-fermentation, rice wine with high levels of resveratrol (86 mg/L) was obtained. Ultrafiltration was also applied as an alternative treatment to boiling for the clarification and sterilization of the beverage while maintaining the bioactive components.

In summary, several strategies have been applied to obtain bioactive products from different sources, including agro-industrial byproducts. We are pleased to present this special issue, which offers new opportunities for alternative strategies in food production, supply, and consumption to move into a more sustainable industry with zero-waste. However, further work is required to test their food applicability, for example, as antioxidant additives, while functional ingredients of particular importance are performing more in vivo and clinical studies to demonstrate bioactivity and functionality. Nonetheless, some of the products and extracts are based on phenolic compounds, whose bioactivity is highly recognized.

Acknowledgments: The authors gratefully acknowledge the postdoctoral grant funded by the "Acción 6 del Plan de Apoyo a la Investigación de la Universidad de Jaén (UJA), 2017–2019" (Spain) and FEDER UJA project 1260905 funded by "Programa Operativo FEDER 2014–2020" and "Consejería de Economía y Conocimiento de la Junta de Andalucía" (Spain). Finally, we would like to thank the authors for their contributions and the valuable work of the reviewers in the evaluation process.

Conflicts of Interest: The authors declare no conflict of interest.

References

1. Sharifi-Rad, M.; Ozcelik, B.; Altın, G.; Daşkaya-Dikmen, C.; Martorell, M.; Ramírez-Alarcón, K.; Alarcón-Zapata, P.; Morais-Braga, M.F.B.; Carneiro, J.N.P.; Borges Leal, A.L.A.; et al. plants-from farm to food applications and phytopharmacotherapy. *Trends Food Sci. Technol.* **2018**, *80*, 242–263. [CrossRef]
2. Contreras, M.d.M.; Gómez-Sala, B.; Martín-Álvarez, P.J.; Amigo, L.; Ramos, M.; Recio, I. Monitoring the large-scale production of the antihypertensive peptides RYLGY and AYFYPEL by HPLC-MS. *Anal. Bioanal. Chem.* **2010**, *397*, 2825–2832. [CrossRef] [PubMed]
3. Salehi, B.; Abu-Darwish, M.S.; Tarawneh, A.H.; Cabral, C.; Gadetskaya, A.V.; Salgueiro, L.; Hosseinabadi, T.; Rajabi, S.; Chanda, W.; Sharifi-Rad, M.; et al. Plants—Food applications and phytopharmacy properties. *Trends Food Sci. Technol.* **2019**, *85*, 287–306. [CrossRef]
4. Li, P.H.; Lu, W.C.; Chan, Y.J.; Zhao, Y.P.; Nie, X.B.; Jiang, C.X.; Ji, Y.X. Feasibility of using seaweed (*Gracilaria coronopifolia*) synbiotic as a bioactive material for intestinal health. *Foods* **2019**, *8*, 623. [CrossRef] [PubMed]
5. Contreras, M.d.M.; Lama, A.; Gutiérrez-Pérez, J.; Espínola, F.; Moya, M.; Castro, E. Protein extraction from agri-food residues for integration in biorefinery: Potential techniques and current status. *Bioresour. Technol.* **2019**, *280*, 459–477. [CrossRef] [PubMed]
6. Llerena, W.; Samaniego, I.; Angós, I.; Brito, B.; Ortiz, B.; Carrillo, W. Biocompounds content prediction in ecuadorian fruits using a mathematical model. *Foods* **2019**, *8*, 284. [CrossRef] [PubMed]
7. Martín-García, B.; Pasini, F.; Verardo, V.; Gómez-Caravaca, A.M.; Marconi, E.; Caboni, M.F. Distribution of free and bound phenolic compounds in buckwheat milling fractions. *Foods* **2019**, *8*, 670. [CrossRef] [PubMed]
8. Mekky, R.H.; Abdel-Sattar, E.; Segura-Carretero, A.; Contreras, M.D.M. Phenolic compounds from sesame cake and antioxidant activity: A new insight for agri-food residues' significance for sustainable development. *Foods* **2019**, *8*, 432. [CrossRef] [PubMed]

9.   Contreras, M.D.M.; Lama-Muñoz, A.; Gutiérrez-Pérez, J.M.; Espínola, F.; Moya, M.; Romero, I.; Castro, E. Integrated process for sequential extraction of bioactive phenolic compounds and proteins from mill and field olive leaves and effects on the lignocellulosic profile. *Foods* **2019**, *8*, 531. [CrossRef] [PubMed]

10.   Yang, K.-R.; Yu, H.-C.; Huang, C.-Y.; Kuo, J.-M.; Chang, C.; Shieh, C.-J.; Kuo, C.-H. Bioprocessed production of resveratrol-enriched rice wine: Simultaneous rice wine fermentation, extraction, and transformation of piceid to resveratrol from *Polygonum cuspidatum* roots. *Foods* **2019**, *8*, 258. [CrossRef] [PubMed]

Article

Distribution of Free and Bound Phenolic Compounds in Buckwheat Milling Fractions

Beatriz Martín-García [1], Federica Pasini [2], Vito Verardo [3,4,*], Ana María Gómez-Caravaca [1], Emanuele Marconi [5] and Maria Fiorenza Caboni [2,6]

[1] Department of Analytical Chemistry, Faculty of Sciences, University of Granada, Avd. Fuentenueva s/n, 18071 Granada, Spain; bearu15@correo.ugr.es (B.M.-G.); anagomez@ugr.es (A.M.G.-C.)
[2] Department of Agricultural and Food Sciences, University of Bologna, Piazza Goidanich 60, (FC) 47521 Cesena, Italy; federica.pasini5@unibo.it (F.P.); maria.caboni@unibo.it (M.F.C.)
[3] Department of Nutrition and Food Science, University of Granada, Campus of Cartuja, 18071 Granada, Spain
[4] Institute of Nutrition and Food Technology 'José Mataix', Biomedical Research Center, University of Granada, Avda del Conocimiento sn., 18100 Armilla, Granada, Spain
[5] Dipartimento Agricoltura, Ambiente e Alimenti, Università del Molise, via De Sanctis s/n, I-86100 Campobasso, Italy; marconi@unimol.it
[6] Interdepartmental Centre for Agri-Food Industrial Research, Alma Mater Studiorum, Università di Bologna, via Quinto Bucci 336, 47521 Cesena (FC), Italy
* Correspondence: vitoverardo@ugr.es; Tel.: +34-958243863

Received: 16 October 2019; Accepted: 9 December 2019; Published: 12 December 2019

Abstract: Buckwheat is a rich source of phenolic compounds that have shown to possess beneficial effect to reduce some diseases due to their antioxidant power. Phenolic compounds are present in the free and in the bound form to the cell wall that are concentrated mainly in the outer layer (hull and bran). Hull is removed before the milling of buckwheat to obtain flours. In order to evaluate the phenolic composition in dehulled buckwheat milling fractions, it was carried out a determination of free and bound phenolic compounds in dehulled whole buckwheat flour, light flour, bran flour, and middling flour by high-performance liquid chromatography-mass spectrometry (HPLC–MS). The most abundant free phenolic compounds were rutin and epiafzelchin–epicatechin-O-dimethylgallate, whereas the most abundant bound phenolic compounds were catechin and epicatechin in all buckwheat flours. Besides, the highest content of free phenolic compounds was obtained in bran flour (1249.49 mg/kg d.w.), whereas the greatest bound phenolic content was in middling (704.47 mg/kg d.w.) and bran flours (689.81 mg/kg d.w.). Thus, middling and bran flours are naturally enriched flours in phenolic compounds that could be used to develop functional foods.

Keywords: free and bound phenolic compounds; buckwheat flours; HPLC–MS; milling fractions

---

1. Introduction

Buckwheat (*Fagopyrum esculentum* Moench) as a traditional pseudocereal crop which belongs to the Polygonaceae is extensively utilized as food and as a medicinal plant [1]. Buckwheat is a rich source of starch, protein, and vitamins [2]. In addition, buckwheat is well known for containing phenolic compounds, including phenolic acids such as protocatechuic, syringic acid, and caffeic acid and flavonoids such as rutin (quercetin 3-rutinoside), quercetin, hyperoside (quercetin 3-O-b-D-galactoside), quercitrin (quercetin 3-O-a-L-rhamnoside), epicatechin, orientin, vitexin, isovitexin, and isoorientin [3–5]. Rutin is the most concentrated phenolic compound in Tartary and some common buckwheats, which have a content higher than most other plants [2]. Phenolic compounds in buckwheat have shown to possess antioxidant activity which has been associated with a lower incidence of cardiovascular disease, cancers, and age-related degenerative process [6–10].

Phenolic compounds in buckwheat are present in the free and in the bound form to cell wall [11], however, the majority of phenolic compounds are present in the free form, which has a distribution and concentration that is different in each part of the grain: pericarp (hull, husk), coat, endosperm, embryo with axis, and two cotyledons [12]; phenolic compounds are concentrated in the outer layers (hull and bran) of buckwheat grain [2]. Nevertheless, during buckwheat seeds processing into flour, the hull (17–20% of buckwheat grain) is removed by stone dehuller. The resulting product, called groat (intact achene), is milled into bran flour (10–24%), which is a by-product that it is not commonly used in foods, and light flour (55–70%), which consists principally of endosperm and is used in human nutrition [13]. In addition, middling is a by-product from buckwheat milling that is not a flour that comprises different fractions and it includes 12% of the original grain, consisting of fractions of endosperm, bran, and germ [14]. Milling techniques used in the food industry employ mechanical force to break the grains into smaller fragments or fine particles. [15]. Previous studies reported the use of roller milling process in dehulled whole buckwheat to obtain a flour and the separation of this flour into various fractions from outer to inner parts [2,16]. These studies have shown that outer layers are richer in protein, lipid, dietary fiber, and ash content than the inner layers. Also, the antioxidant capacity in flour fractions in the outer layers is higher than that in the inner layers by the increase of phenolic compounds from bran [2,16]. In addition, it has reported that milling fractions that contain outer layers possess a higher concentration of phenolic compounds than whole grain and groat flour fractions [6]. Therefore, the aim of this work was the determination of free and bound phenolic content in different buckwheat meals/flours: whole grain flour, light flour, bran meals, and middling flour in order to evaluate the phenolic concentration in each buckwheat meal fraction. These analyses will furnish new information about the total content of phenolic compounds in each fraction, taking into account the free or extractable fraction and bound or nonextractable phenolic fraction (NEPP). For that purpose, phenolic compounds were extracted and then were analyzed by high-performance liquid chromatography-mass spectrometry (HPLC-MS).

2. Materials and Methods

2.1. Sample

Buckwheat meals/flours were obtained from whole buckwheat grain (cv. Darja) harvested in Matrice (Italy) (41°37′00″ N 14°43′00″ E), situated in a hilly location at 750 m above sea level. The field presented high tenacity of the soil due to the presence of clay. Harvesting took place on September 2018. The grain was dehulled by stone dehuller (GRANO 200 SCHNITZER Stein-Getreidemuhle, Offenburg, Germany), and the groat (dehulled grain) was roller-milled by using an experimental mill (Labormill 4RB Bona, Monza, Italy). This mill is able to produce three milling fractions with different particle sizes that constituted the basis for differentiation between bran meal, middling flour, and light flour (Figure 1). In the bran meal, the majority of particles were >505 μm, while in middling flour, between 219–363 μm, and in light flour, <183 μm. Granulometry analysis was performed using an automatic sieve (Buhler ML1-300, Uzwil, Switzerland).

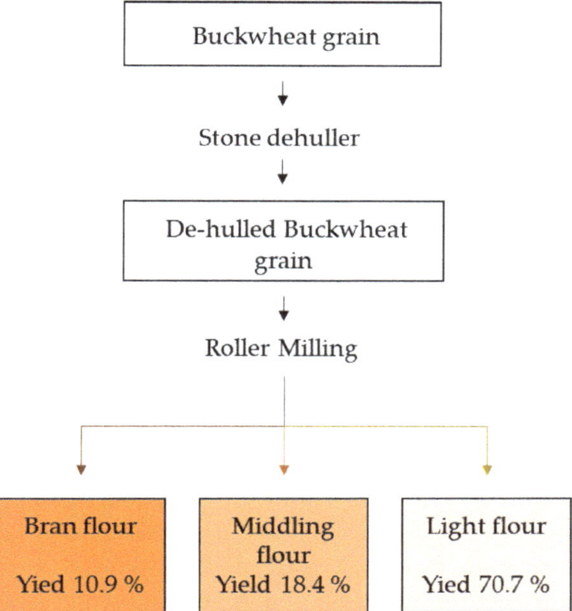

Figure 1. Flowchart of the milling process used for the production of buckwheat flours.

2.2. Reagents and Chemicals

HPLC-grade acetonitrile, water, methanol, acetone, acetic acid, ethanol, hexane, ethyl acetate, diethyl ether, and hydrochloric acid were purchased from Merck KGaA (Darmstadt, Germany). Hydroxide sodium was from Fluka (Buchs, Switzerland). Ferulic acid, catechin, quercetin, and rutin (Sigma-Aldrich (St. Louis, MO, USA)) were used for the calibration curves.

2.3. Extraction Method

Extraction of free phenolic compounds from buckwheat flour fractions has been carried out according with the method established by Hung & Morita (2008) [2] with certain modifications. One gram of buckwheat flour was extracted thrice in an ultrasonic bath with a solution of ethanol/water (4:1 v/v). The supernatants were collected, centrifugated at 2500 rpm for 10 minutes, evaporated, and reconstituted with 1 mL of methanol/water (1:1 v/v). The extracts were stored at $-18\,°C$ until use.

Extraction of bound phenolic compounds was carried out according to the method established by Verardo et al. (2011) [5]: residues of free phenolic extraction were digested with 25 mL of 1M NaOH at room temperature for 18 h by shaking under nitrogen gas. The mixture was acidified (pH 2.2–2.5) with hydrochloric acid in a cooling ice bath and extracted with 250 mL of hexane to remove the lipids. The aqueous solution was extracted five times with 50 mL of 1:1 diethyl ether/ethyl acetate (v/v). The organic fractions were collected and evaporated at 40 °C in a rotary evaporator. The dry extract was reconstituted in 1 mL of methanol/water (1:1 v/v) and stored at $-18\,°C$ until use.

2.4. Determination of Free and Bound Phenolic Compounds by HPLC–MS

A liquid chromatography apparatus HP 1100 Series (Agilent Technologies, Palo Alto, CA, USA) equipped with a degasser, a binary pump delivery system, and an automatic liquid sampler and coupled to a single quadrupole mass spectrometer detector was used. Separation of free and bound phenolic compounds from buckwheat flour fractions was carried out using a C-18 column (Poroshell

120, SB-C18, 3.0 × 100 mm, 2.7 μm from Agilent Technologies, Palo Alto, CA, USA). The gradient elution was the same as that previously established by Gómez-Caravaca et al. [17] using as a mobile phase A acidified water (1% acetic acid) and as mobile phase B acetonitrile. MS analysis was carried out using an electrospray ionization (ESI) interface in negative ionization mode at the following conditions: drying gas flow (N_2), 9.0 L/min; nebulizer pressure, 50 psi; gas drying temperature, 350 °C; capillary voltage, 4000 V. The fragmentor and m/z range used for HPLC–ESI/MS analyses were 80 V and m/z 50–1000, respectively. Data were processed by the software MassHunter Workstation Qualitative Analysis Version B.07.00 (Agilent Technologies, Santa Clara, CA, USA).

2.5. Statistical Analysis

The results of quantification reported in this work are the averages of three repetitions ($n = 3$). Tukey's honest significant difference multiple comparison (one-way ANOVA) at the $p < 0.05$ level was evaluated by using the Statistica 7.0 software (StatSoft, Tulsa, OK, USA)

3. Results and Discussion

3.1. Analytical Parameters of the Method

An analytical validation of the method was performed considering linearity and sensitivity. In order to quantify phenolic compounds in buckwheat fractions, five calibrations curves were elaborated with the standards ferulic acid, catechin, quercetin, gallic acid, and rutin. Table 1 includes the analytical parameters of the standards used, containing calibration ranges, calibration curves, determination coefficients, limit of detection (LOD), and limit of quantification (LOQ).

Table 1. Analytical parameters of the method proposed.

Standards	Calibration Ranges (mg/L)	Calibration Curves (mg/g)	r^2	LOD (mg/L)	LOQ (mg/L)
Ferulic acid	LOQ-100	$y = 119572x + 16157$	0.9995	0.0136	0.0452
Catechin	LOQ-100	$y = 170925x + 8609.5$	0.9994	0.0095	0.0316
Quercetin	LOQ-100	$y = 402162x + 44862$	0.9996	0.0040	0.0134
Gallic acid	LOQ-100	$y = 123892x - 4971.6$	0.9984	0.0131	0.0437
Rutin	LOQ-100	$y = 199694x - 2067.2$	0.9999	0.0081	0.0271

LOD: Limit of detection, LOQ: Limit of quantification.

Calibration curves were carried out by using the peak areas analyte standard against the concentration of the analyte for the analysis by HPLC. The external calibration of the standards was elaborated at different concentration levels from LOQ to 100 mg L^{-1}. All calibration curves revealed good linearity among different concentrations, and the determination coefficients were higher than 0.9994 in all cases. The method used for analysis showed LOD within the range 0.0040–0.0136 mg L^{-1}, the LOQ were within 0.0134–0.0452 mg L^{-1}.

3.2. Identification of Phenolic Compounds from Buckwheat Extracts by HPLC–MS

Free and bound phenolic compounds in buckwheat flour fractions extracts were analyzed by HPLC with MS detection, and the identification of these compounds was carried out by comparison of molecular weight in bibliography and when available, by co-elution with commercial standards.

A total of 25 free phenolic compounds were identified in buckwheat flours, among them five were phenolic acids and 20 were flavonoids, and they were previously identified in other works [4,18] (Table 2).

Table 2. Identification table of free phenolic compounds in buckwheat flours.

Peak	Retention Time	[M-H]	Molecular Formula	Compound	In Source Fragments
1	2.1	315	$C_{13}H_{15}O_9$	2-hydroxy-3- O-β-D-glucopyranosylbenzoic acid	153
2	2.6	315	$C_{13}H_{15}O_9$	Protocatechuic-4-O-glucoside acid	153
3	3.3	451	$C_{21}H_{23}O_{11}$	Catechin-glucoside	289
4	4.1	341	$C_{15}H_{17}O_9$	Caffeic acid hexose	179
5	4.2	289	$C_{15}H_{13}O_6$	Catechin	
6	4.4	487	$C_{21}H_{27}O_{13}$	Swertiamacroside	179
7	5.0	179	$C_9H_7O_4$	Caffeic acid	
8	5.5	289	$C_{15}H_{13}O_6$	Epicatechin	
9	6.2	561	$C_{30}H_{25}O_{11}$	(Epi)afzelchin-(epi) catechin isomer A	543, 289, 271, 435
10	6.8	447	$C_{21}H_{19}O_{11}$	Orientin	
11	7.0	447	$C_{21}H_{19}O_{12}$	Isorientin	
12	7.8	431	$C_{21}H_{19}O_{10}$	Vitexin	
13	7.9	609	$C_{27}H_{29}O_{16}$	Rutin	
14	7.9	441	$C_{22}H_{17}O_{10}$	Epicatechin-gallate	289
15	8.0	833	$C_{45}H_{37}O_{16}$	Epiafzelchin–epiafzelchin–epicatechin	
16	8.2	487	$C_{21}H_{27}O_{13}$	Swertiamacroside	
17	8.3	463	$C_{21}H_{19}O_{12}$	Hyperin	
18	8.7	727	$C_{38}H_{31}O_{15}$	Epiafzelchin-epicatechin-O-methylgallate	455, 289, 271
19	9.4	455	$C_{23}H_{19}O_{10}$	(−)-Epicatechin-3-(3″-O-methyl) gallate	289
20	9.5	561	$C_{30}H_{25}O_{11}$	(Epi)afzelchin-(epi)catechin isomer B	543, 425, 289, 271
21	9.9	757	$C_{39}H_{33}O_{16}$	Procyanidin B2-dimethylgallate	
22	10.7	741	$C_{39}H_{33}O_{15}$	Epiafzelchin–epicatechin-O-dimethylgallate	
23	11.5	469	$C_{24}H_{21}O_{10}$	Epicatechin-O-3,4-dimethylgallate	
24	12.3	463	$C_{21}H_{19}O_{12}$	Isoquercitrin	
25	12.6	301	$C_{15}H_{10}O_7$	Quercetin	

Twenty-four bound phenolic compounds were identified in buckwheat flours: seven were phenolic acid derivatives and 17 were flavonoids, which were identified in previous works (Table 3) [5,18].

Table 3. Identification of bound phenolic compounds in buckwheat flours.

Peak	Retention Time	[M-H]	Molecular Formula	Compound
1	2.1	315	$C_{13}H_{15}O_9$	2-hydroxy-3-O-β-D-glucopyranosylbenzoic acid
2	2.6	315	$C_{13}H_{15}O_9$	Protocatechuic-4-O-glucoside acid
3	3.2	341	$C_{15}H_{17}O_9$	Caffeic acid hexose isomer a
4	4.1	341	$C_{15}H_{17}O_9$	Caffeic acid hexose isomer b
5	4.2	289	$C_{15}H_{13}O_6$	Catechin
6	4.4	487	$C_{21}H_{27}O_{13}$	Swertiamacroside isomer a
7	5.0	179	$C_9H_7O_4$	Caffeic acid
8	5.5	289	$C_{15}H_{13}O_6$	Epicatechin
9	6.3	197	$C_9H_9O_5$	Syringic acid
10	6.8	447	$C_{21}H_{19}O_{11}$	Orientin
11	6.9	163	$C_9H_7O_3$	*p*-coumaric acid derivative
12	7.0	575	$C_{30}H_{23}O_{12}$	Procyanidin A
13	7.5	317	$C_{15}H_9O_8$	Myricetin
14	7.8	431	$C_{21}H_{19}O_{10}$	Vitexin
15	7.9	609	$C_{27}H_{29}O_{16}$	Rutin
16	7.9	441	$C_{22}H_{17}O_{10}$	Epicatechin gallate
17	8.2	451	$C_{21}H_{23}O_{11}$	Catechin-glucoside
18	8.2	487	$C_{21}H_{27}O_{13}$	Swertiamacroside isomer b
19	8.7	727	$C_{38}H_{31}O_{15}$	Epiafzelchin–epicatechin-O-methylgallate
20	9.3	163	$C_9H_7O_3$	*p*-coumaric acid
21	9.4	455	$C_{23}H_{19}O_{10}$	(−)-epicatechin-3-(3″-O-methyl) gallate
22	11.5	469	$C_{24}H_{21}O_{10}$	Epicatechin-O-3,4-dimethylgallate
23	12.3	463	$C_{21}H_{19}O_{12}$	Isoquercitrin
24	12.6	301	$C_{15}H_{10}O_7$	Quercetin

3.3. Quantification of Free and Bound Phenolic Compounds in Buckwheat Fractions

Free phenolic compounds were quantified through of calibration curves of standards. A total of 25 free phenolic compounds were quantified in buckwheat meals/flours: de-hulled grain meal, bran meal, middling flour, and light flour (Table 4).

Table 4. Free phenolic compounds quantified in buckwheat meals/flours (mg/kg d.w.) determined by HPLC-MS.

Free Phenolic Compounds	Bran Meal	Middling Flour	Light Flour	De-hulled Grain Meal
2-hydroxy-3-*O*-β-Dglucopyranosylbenzoic acid	42.17 [b]	78.22 [a]	2.67 [d]	32.71 [c]
Protocatechuic-4-*O*-glucoside acid	79.69 [b]	120.59 [a]	2.93 [d]	65.56 [c]
Catechin-glucoside	23.87 [b]	34.97 [a]	1.88 [d]	13.53 [c]
Caffeic acid hexose	41.02 [a]	37.39 [b]	1.06 [d]	30.95 [c]
Catechin	20.40 [a]	17.25 [b]	1.36 [d]	7.33 [c]
Swertiamacroside	33.14 [a]	22.81 [b]	0.85 [d]	9.84 [c]
Caffeic acid	36.82 [a]	22.35 [b]	0.15 [d]	0.96 [c]
Epicatechin	69.56 [a]	26.48 [b]	2.60 [d]	14.01 [c]
(Epi)afzelchin-(epi) catechin isomer A	58.11 [a]	35.49 [b]	1.71 [d]	20.06 [c]
Orientin	5.18 [a]	3.79 [b]	0.02 [d]	1.58 [c]
Isorientin	4.61 [a]	2.84 [b]	<LOQ	0.82 [c]
Vitexin	9.14 [a]	6.26 [b]	0.06 [d]	2.02 [c]
Rutin	214.99 [a]	148.63 [b]	7.03 [d]	87.33 [c]
Epicatechin-gallate	18.56 [a]	7.82 [b]	0.28 [d]	5.22 [c]
Epiafzelchin–epiafzelchin–epicatechin	20.37 [a]	12.69 [b]	0.84 [d]	8.01 [c]
Swertiamacroside	27.41 [a]	20.92 [b]	4.23 [d]	9.47 [c]
Hyperin	2.84 [a]	1.59 [b]	<LOQ	0.13 [c]
Epiafzelchin–epicatechin-*O*-methyl gallate	76.84 [a]	39.84 [b]	1.00 [d]	28.73 [c]
(−)-Epicatechin-3-(3″-*O*-methyl) gallate	31.61 [a]	17.77 [b]	0.51 [d]	15.18 [c]
(Epi)afzelchin-(epi) catechin isomer B	25.04 [a]	15.03 [b]	0.47 [d]	9.95 [c]
Procyanidin B2-dimethylgallate	51.46 [a]	29.22 [b]	0.67 [d]	21.06 [c]
Epiafzelchin–epicatechin-*O*-dimethylgallate	216.94 [a]	176.67 [b]	13.11 [d]	93.83 [c]
Epicatechin-*O*-3,4-dimethylgallate	98.07 [a]	8.05 [c]	2.31 [d]	39.10 [b]
Isoquercitrin	1.41 [a,b]	2.05 [a]	0.54 [d]	1.09 [c]
Quercetin	33.21 [a]	12.39 [b]	0.06 [d]	2.27 [c]
Flavonoids	982.23 [a]	598.23 [b]	34.47 [d]	371.25 [c]
Phenolic acids	260.26 [b]	302.28 [a]	11.89 [d]	149.49 [c]
Sum	1242.49 [a]	901.10 [b]	46.36 [d]	520.74 [c]

Different letters in the same line show significant differences ($p < 0.05$), LOQ: Limit of quantification.

The most concentrated free phenolic compound in all buckwheat flours was epiafzelchin–epicatechin-*O*-dimethylgallate, whose content was 13.11 mg/kg d.w. in light flour, 93.83 mg/kg d.w. in de-hulled grain meal, 176.67 mg/kg d.w. in middling flour, and 216.94 mg/kg d.w. in bran meal. The second most concentrated in buckwheat flours was rutin, whose content was from 7.03 mg/kg d.w in light flour, 87.33 mg/kg d.w. in de-hulled grain meal, 148.63 mg/kg d.w. in middling flour, to 214.99 mg/kg d.w in bran meal. Thus, the most abundant free flavonoids are present in buckwheat bran meal, followed by middling flour, de-hulled buckwheat meal, and light flour. Besides, 2-hydroxy-3-*O*-β-D-glucopyranosylbenzoic and protocatechuic-4-*O*-glucoside acid appear in buckwheat fractions in significant quantities, whose values were 2.67–2.93 mg/kg d.w. in light flour, 32.71–65.56 mg/kg in de-hulled grain meal, 42.17–79.69 mg/kg d.w. in bran meal, and 78.22–120.56 mg/kg d.w. in middling flour. Therefore, the highest content of phenolic acids appears in middling flour, followed by bran meal, de-hulled grain meal, and light flour. The third most abundant phenolic compound in middling and de-hulled grain meal was protocatechuic-4-*O*-glucoside acid (120.59 and 65.56 mg/kg d.w.), whereas in light flour was swertiamacroside (4.23 mg/kg d.w.), and in bran meal was epicatechin-*O*-3,4-dimethylgallate (98.07 mg/kg d.w.).

The total free phenolic content in buckwheat flours was decreasing in the following order: bran meal > middling flour > de-hulled buckwheat meal > light flour (1242.49, 901.10, 520.74, and 46.36 mg/kg d.w.). These results are due to the most abundant free phenolic compounds being flavonoids, which corresponded to 66–79% of total free phenolic compounds, and these are found in higher concentration in outer layers than in inner layers of buckwheat grain [2]. For that reason, bran meal contains the highest content of free phenolic compounds, followed by middling flour, as it contains seed coat.

The concentration of free phenolic compounds obtained in buckwheat was compared with that obtained previously in other works. Verardo et al. (2011) [5] quantified the individual free phenolic compounds in de-hulled buckwheat grain, where rutin was the most concentrated, whose value was 35.12% higher than that obtained in the present work and total content of free phenolic compounds was 48.39% higher than in the present work. Nevertheless, the most concentrated free phenolic compound in our work was epiafzelchin–epicatechin-*O*-dimethylgallate, whose value was 50% higher than that obtained by Verardo et al. (2011) [5]. These differences of concentration could be due to the different buckwheat cultivar. Besides, Inglett et al. (2011) [18] quantified the free flavonoid content in different buckwheat flours (fancy, farinetta, supreme, and whole), fancy corresponded with light flour, supreme flour is similar to bran meal, farinetta consists of a fine granulated mixture of aleurone layer of hulled achene and achene embryo, a composition similar to middling flour [19,20]. The value of free flavonoids obtained in our study in light flour, de-hulled grain meal, bran meal, and middling flour (34.47 mg/kg d.w., 371.25 mg/kg d.w., 982.23 mg/kg d.w., and 598.83 mg/kg d.w) were in the same order of magnitude than that obtained in fancy (71.40 mg/kg d.w.), whole buckwheat flour (417.03 mg/kg d.w.), supreme (525.27 mg/kg d.w.), and farinetta (671.50 mg/kg d.w.) by Inglett et al. (2011) [18].

Hung et al. (2008) [2] reported the content of rutin in the free form obtained in different buckwheat flour fractions, and its concentration was 2.5–3 mg/kg d.w. in the innermost layers, whereas in the outer layers, it was 274-337.8 mg/kg. These results were similar to those obtained in the present work in the light flour (7.03 mg/kg dw.) and bran meal (214.99 mg/kg d.w.). Kalinová et al. (2019) [21] reported the free phenolic compounds in the seed coat (553.18 mg/kg d.w.), in the endosperm (2.59 mg/kg d.w.), and in the groat (139.66 mg/kg d.w.). These values were lower than those obtained in bran meal, light flour, and de-hulled grain meal, and also, the content of rutin in seed coat (54.23 mg/kg d.w.) represents a quart of the phenolic bran meal (214.99 mg/kg d.w.) obtained in our study. This could be due to the different cultivar and/or the different methodology of determination of phenolic compounds (by MS detection a higher number of compounds are determined).In addition, Liu et al. (2019) [22] reported the concentration of rutin in common buckwheat (62.19 mg/kg d.w.) that was in the same order as that obtained in de-hulled grain meal (87.33 mg/kg d.w.). According to the results obtained in these previous works, it has shown that rutin in the free form is concentrated in the outer layers, which is in concordance with our results.

The Table 5 reports the content of bound phenolic compounds in buckwheat flours. Bound phenolic compounds composition in buckwheat flours was similar than that obtained in free phenolic fraction; nevertheless, flavonoids such as isorientin, epiafzelchin–epiafzelchin–epicatechin, Procyanidin B2-dimethylgallate, hyperin, and (epi)afzelchin-(epi)catechin were not detected in bound fraction, whereas some phenolic acids such as syringic and p-coumaric acid, procyanidin A, and myricetin were determined only in bound fraction.

Catechin was the most concentrated bound phenolic compound in all buckwheat flours, representing 25–30% of total bound phenolic compounds, and its concentration was 54.67 mg/kg d.w. in light flour, 95.45 mg/kg d.w. in de-hulled grain meal, 200.17 mg/kg d.w. in middling flour, and 207.74 mg/kg d.w. in bran meal, respectively. The second component most abundant was epicatechin, whose content was 34.67 mg/kg d.w. in light flour, 41.55 mg/kg d.w. in de-hulled grain meal, 59.08 mg/kg d.w. in bran flour, and 97.50 mg/kg d.w. in middling flour. The third most abundant phenolic compound in de-hulled grain meal and bran meal was syringic acid (35.62 mg/kg d.w. and 85.86 mg/kg d.w.), whereas in middling flour it was caffeic acid hexose (56.73 mg/kg d.w.), and in light flour it was swertiamacroside.

The total bound phenolic content in buckwheat flours was increasing in the following order: light flour < de-hulled grain meal < bran meal < middling flour (207.74, 389.51, 689.81, and 704.47 mg/kg d.w.). Therefore, the highest concentration of bound phenolic compounds is in middling and bran meal due to these compounds being linked to the cell wall of buckwheat layers. Flavonoids represented 59–65% of the bound phenolic fraction. Whereas, phenolic acids represented 35–41% of bound phenolic fraction.

Concentrations of catechin, epicatechin, syringic, and total bound phenolic compounds in de-hulled whole buckwheat flour obtained by Verardo et al. (2011) [5] were 23.88%, 48.54%, and 53.18% higher than those obtained in the present work. Inglett et al. (2011) [18] reported the content of total bound flavonoid in buckwheat flour fractions obtained was 59.25 mg/kg d.w. in fancy, 389.68 mg/kg in farinetta, 530.21 mg/kg in supreme, and 613.77 mg/kg d.w. in whole flour, which are in the same order of magnitude as that obtained in our work. Nevertheless, in this study, the highest bound phenolic content was obtained in whole buckwheat flour, whereas in our work, the maximum value of phenolic content corresponded with the middling flour. This could be due to the different cultivar or because Inglett et al. (2011) [18] could include the hull in the buckwheat grain flour.

Table 5. Bound phenolic compounds quantified in buckwheat meals/flours (mg/kg d.w.) determined by HPLC–MS.

Bound Phenolic Compounds	Bran Meal	Middling Flour	Light Flour	De-Hulled Grain Meal
2-hydroxy-3-O-β-D-glucopyranosylbenzoic acid	23.02 [b]	34.56 [a]	6.19 [c,d]	7.88 [d]
Protocatechuic-4-O-glucoside acid	18.44 [b]	25.50 [a]	5.51 [c]	5.95 [c]
Caffeic acid hexose isomer a	5.52 [b]	11.34 [a]	0.67 [c]	0.43 [c,d]
Caffeic acid hexose isomer b	40.42 [b]	56.73 [a]	13.28 [d]	26.35 [c]
Catechin	207.74 [a]	200.17 [a]	54.67 [c]	95.45 [b]
Swertiamacroside	23.25 [c,d]	31.84 [a,b]	25.40 [d]	33.66 [a]
Caffeic acid	<LOQ	<LOQ	<LOQ	<LOQ
Epicatechin	59.08 [b]	97.50 [a]	34.67 [d]	41.55 [c]
Syringic acid	85.86 [a]	43.57 [b]	7.74 [d]	35.62 [c]
Orientin	0.46 [a]	0.56 [a]	0.19 [c]	0.22 [b]
p-coumaric acid derivative	9.65 [a]	3.53 [b]	1.39 [d]	3.24 [c]
Procyanidin A	8.82 [a]	9.03 [a]	0.95 [c]	4.95 [b]
Myricetin	4.12 [a]	3.80 [a]	2.06 [b,c]	2.92 [b]
Vitexin	4.22 [a]	3.86 [a]	0.67 [c]	2.30 [b]
Rutin	51.64 [a]	45.19 [b]	6.82 [d]	33.71 [c]
Epicatechin gallate	16.24 [a]	15.57 [a]	4.21 [c]	10.75 [b]
Catechin-glucoside	16.48 [a]	17.51 [a]	1.04 [c]	13.26 [b]
Swertiamacroside	39.40 [a]	32.37 [b]	23.52 [d]	30.43 [c]
Epiafzelchin–epicatechin-O-methylgallate	28.04 [a]	27.81 [a]	3.57 [c]	9.72 [b]
p-coumaric acid	3.96 b	6.91 [a]	0.67 [d]	2.74 [c]
(−)-epicatechin-3-(3″-O-methyl) gallate	6.09 [a]	6.05 [a]	2.06 [c]	4.17 [b]
Epicatechin-O-3,4-dimethylgallate	4.65 [a]	4.11 [a]	0.50 [c]	1.78 [b]
Isoquercitrin	6.06 [a]	5.89 [a]	1.03 [c]	3.64 [b]
Quercitrin	26.64 [a]	21.05 [b]	10.94 [d]	18.78 [c]
Flavonoids	440.29 [b]	458.11 [a]	123.37 [d]	243.20 [c]
Phenolic acids	249.52 [a]	246.35 [b]	84.37 [d]	146.31 [c]
Total	689.81 [b]	704.47 [a]	207.74 [d]	389.51 [c]

Different letters in the same line show significant differences ($p < 0.05$), LOQ: Limit of quantification.

The total content of flavonoids was from 157.84 mg/kg d.w. in light flour to 1422.52 mg/kg d.w. in bran meal, whereas the content of phenolic acids was from 96.261 mg/kg d.w. in light flour to 548.63 mg/kg d.w. in middling flour. Total phenolic content was from 254.10 mg/kg d.w. in light flour to 1932.30 mg/kg d.w. in bran meal (Table 6). According to the results, the total phenolic content was increasing in the following order: light flour < de-hulled grain meal < middling flour < bran meal Therefore, middling flour and bran meal possess the highest phenolic content due to bran and the aleurone layer being richer in many phenolic compounds than the others buckwheat flours [23]. Total flavonoid obtained in de-hulled grain meal, bran meal, and middling flour was 49.22%, 71.21%, and 27.83% higher than that obtained in whole grain meal, supreme, and farinetta by liquid chromatography-electrospray ionization- mass spectrometry (LC–ESI-MS) [18]. According to Guo and co-workers, free rutin was determined in a range of 51–81% [24].

Table 6. Total content of flavonoids, phenolic acids, and phenolic compounds in buckwheat flours. Results are expressed as mg/kg d.w.

	Flavonoids	Phenolic Acids	Total
Bran meal	1422.52 [a]	509.78 [b]	1932.30 [a]
Middling flour	1056.94 [b]	548.63 [a]	1605.57 [b]
Light flour	157.84 [d]	96.261 [d]	254.10 [d]
De-hulled grain meal	614.46 [c]	295.80 [c]	910.25 [c]

Different letters in the same column show significant differences ($p < 0.05$).

4. Conclusions

An HPLC–MS has been used for the determination of free and bound phenolic compounds in buckwheat flours: middling flour, bran meal, light flour, and whole meal. The results of this study have shown that the total free phenolic compounds are found in the highest concentration in bran meal, whereas the bound content of phenolic compounds are concentrated in middling flour and bran meal. In buckwheat flours, the main flavonoids were rutin and epiafzelchin–epicatechin-*O*-dimethylgallate, which had the greatest content in bran meal. By contrast, catechin and epicatechin were the main bound flavonoids in buckwheat meal/flours that existed in the greatest quantities in middling and bran fours.

To conclude, the bran meal and middling flour could be considered as flours enriched in phenolic compounds that could be used to elaborate food with health benefits. Moreover, it has been proved, as the distribution of some phenolic compound varied from bran to middling fraction.

Author Contributions: Conceptualization, V.V., E.M., and A.M.G.-C.; Investigation, B.M.-G. and F.P.; Supervision, V.V., A.M.G.-C., and M.F.C.; Writing—original draft, B.M.-G.; Writing—review & editing, F.P., V.V., A.M.G.-C., E.M., and M.F.C.

Funding: This research received no external funding.

Acknowledgments: V.V. thanks the Spanish Ministry of Economy and Competitiveness (MINECO) for "Ramon y Cajal" contract (RYC-2015-18795). B.M.G. would like to thank to the University of Granada "Convocatoria de movilidad internacional de estudiantes de doctorado" grant.

Conflicts of Interest: The authors declare no conflict of interest.

References

1. Sytar, O. Phenolic acids in the inflorescences of different varieties of buckwheat and their antioxidant activity. *J. King Saud Univ.Sci.* **2015**, *27*, 136–142. [CrossRef]
2. Van Hung, P.; Morita, N. Distribution of phenolic compounds in the graded flours milled from whole buckwheat grains and their antioxidant capacities. *Food Chem.* **2008**, *109*, 325–331. [CrossRef] [PubMed]
3. Zhang, Z.; Zhou, M.; Tang, Y.; Li, F.; Tang, Y.; Shao, J. Bioactive compounds in functional buckwheat food. *FRIN* **2012**, *49*, 389–395. [CrossRef]
4. Verardo, V.; Arráez-Román, D.; Segura-Carretero, A.; Marconi, E.; Fernández-Gutiérrez, A.; Caboni, M.F. Identification of buckwheat phenolic compounds by reverse phase high performance liquid chromatography e electrospray ionization-time of flight-mass spectrometry (RP-HPLC e ESI-TOF-MS). *J. Cereal Sci.* **2010**, *52*, 170–176. [CrossRef]
5. Verardo, V.; Arráez-Román, D.; Segura-Carretero, A.; Marconi, E.; Fernández-Gutiérrez, A.; Caboni, M.F. Determination of free and bound phenolic compounds in buckwheat spaghetti by RP-HPLC-ESI-TOF-MS: Effect of thermal processing from farm to fork. *J. Agric. Food Chem.* **2011**, *59*, 7700–7707. [CrossRef]
6. Sedej, I.; Sakač, M.; Mandić, A.; Mišan, A.; Tumbas, V.; Čanadanović-Brunet, J. Buckwheat (Fagopyrum esculentum Moench) Grain and Fractions: Antioxidant Compounds and Activities. *J. Food Sci.* **2012**, *77*, C954–C959. [CrossRef]
7. Kaliora, A.C.; Dedoussis, G.V.Z. Natural antioxidant compounds in risk factors for CVD. *Pharmacol. Res.* **2007**, *56*, 99–109. [CrossRef]
8. Ma, M.S.; In, Y.B.; Hyeon, G.L.; Yang, C.B. Purification and identification of angiotensin I-converting enzyme inhibitory peptide from buckwheat (Fagopyrum esculentum Moench). *Food Chem.* **2006**, *96*, 36–42. [CrossRef]
9. Giménez-Bastida, J.A.; Zieliński, H. Buckwheat as a Functional Food and Its Effects on Health. *J. Agric. Food Chem.* **2015**, *63*, 7896–7913. [CrossRef]
10. Chan, P.K. Inhibition of tumor growth in vitro by the extract of fagopyrum cymosum (fago-c). *Life Sci.* **2003**, *72*, 1851–1858. [CrossRef]
11. Alvarez-Jubete, L.; Wijngaard, H.; Arendt, E.K.; Gallagher, E. Polyphenol composition and in vitro antioxidant activity of amaranth, quinoa buckwheat and wheat as affected by sprouting and baking. *Food Chem.* **2010**, *119*, 770–778. [CrossRef]
12. Bobkov, S. Biochemical and Technological Properties of Buckwheat Grains. In *Molecular Breeding and Nutritional Aspects of Buckwheat*; Zhou, M., Kreft, I., Woo, S.H., Chrungoo, N., Wieslander, G., Eds.; Elsevier: Oxford, UK, 2016; pp. 423–440.
13. Bonafaccia, G.; Marocchini, M.; Kreft, I. Composition and technological properties of the flour. *Food Chem.* **2003**, *80*, 9–15. [CrossRef]

14. Dreher, M. Food sources and uses of dietary fiber. In *Complex Carbohydrates in Foods*; Sungsoo cho, S., Ed.; CRC Press: Boca Raton, FL, USA, 1999; pp. 358–359.

15. Liu, F.; He, C.; Wang, L.; Wang, M. Effect of milling method on the chemical composition and antioxidant capacity of Tartary buckwheat flour. *Int. J. Food Sci. Technol.* **2018**, *53*, 2457–2464. [CrossRef]

16. Morita, N.; Maeda, T.; Sai, R.; Miyake, K.; Yoshioka, H.; Urisu, A.; Adachi, T. Studies on distribution of protein and allergen in graded flours prepared from whole buckwheat grains. *Food Res. Int.* **2006**, *39*, 782–790. [CrossRef]

17. Gómez-Caravaca, A.M.; Verardo, V.; Berardinelli, A.; Marconi, E.; Caboni, M.F. A chemometric approach to determine the phenolic compounds in different barley samples by two different stationary phases: A comparison between C18 and pentafluorophenyl core shell columns. *J. Chromatogr. A* **2014**, *1355*, 134–142. [CrossRef]

18. Inglett, G.E.; Chen, D.; Berhow, M.; Lee, S. Antioxidant activity of commercial buckwheat flours and their free and bound phenolic compositions. *Food Chem.* **2011**, *125*, 923–929. [CrossRef]

19. Inglett, G.E.; Xu, J.; Stevenson, D.G.; Chen, D. Rheological and pasting properties of buckwheat (Fagopyrum esculentum Möench) flours with and without jet-cooking. *Cereal Chem.* **2009**, *86*, 1–6. [CrossRef]

20. Steadman, K.J.; Burgoon, M.S.; Lewis, B.A.; Edwardson, S.E.; Obendorf, R.L. Buckwheat seed milling fractions: Description, macronutrient composition and dietary fibre. *J. Cereal Sci.* **2001**, *33*, 271–278. [CrossRef]

21. Kalinová, J.P.; Vrchotováb, N.; Tříska, J. Phenolics levels in different parts of common buckwheat (Fagopyrum esculentum) achenes. *J. Cereal Sci.* **2019**, *85*, 243–248. [CrossRef]

22. Liu, Y.; Cai, C.; Yao, Y.; Xu, B. Alteration of phenolic profiles and antioxidant capacities of common buckwheat and tartary buckwheat produced in China upon thermal processing. *J. Sci. Food Agric.* **2019**, *99*, 5565–5576. [CrossRef]

23. Fu-Hua, L.; Yuan, Y.; Xiao-lan, Y.; Shu-ying, T.; Jian, M. Phenolic Profiles and Antioxidant Activity of Buckwheat (Fagopyrum esculentum Möench and *Fagopyrum tartaricum* L. Gaerth) Hulls, Brans and Flours. *J. Integr. Agric.* **2013**, *12*, 1684–1693.

24. Guo, X.D.; Wu, C.S.; Ma, Y.J.; Parry, J.; Xu, Y.Y.; Liu, H.; Wang, M. Comparison of milling fractions of tartary buckwheat for their phenolics and antioxidant properties. *Food Res. Int.* **2012**, *49*, 53–59. [CrossRef]

Article

Phenolic Compounds from Sesame Cake and Antioxidant Activity: A New Insight for Agri-Food Residues' Significance for Sustainable Development

Reham Hassan Mekky [1,2], **Essam Abdel-Sattar** [3], **Antonio Segura-Carretero** [2,4,*] **and María del Mar Contreras** [2,4,†]

[1] Department of Pharmacognosy, Faculty of Pharmacy, Egyptian Russian University, Badr City, Cairo-Suez Road, Cairo 11829, Egypt; reham-mekky@eru.edu.eg

[2] Research and Development Functional Food Centre (CIDAF), Bioregión Building, Health Science Technological Park, Avenida del Conocimiento s/n, 18016 Granada, Spain; mcgamez@ujaen.es or mmcontreras@ugr.es or mar.contreras.gamez@gmail.com

[3] Department of Pharmacognosy, Faculty of Pharmacy, Cairo University, Kasr El-Aini Street, Cairo 11562, Egypt; essam.abdelsattar@pharma.cu.edu.eg

[4] Department of Analytical Chemistry, Faculty of Sciences, University of Granada, Avenida Fuentenueva s/n, 18071 Granada, Spain

* Correspondence: ansegura@ugr.es; Tel.: +34-655-984-310

† Present address: Department of Chemical, Environmental and Materials Engineering, Universidad de Jaén, Campus Las Lagunillas, 23071 Jaén, Spain.

Received: 30 August 2019; Accepted: 17 September 2019; Published: 22 September 2019

Abstract: Agri-food residues represent a rich source of nutrients and bioactive secondary metabolites, including phenolic compounds. The effective utilization of these by-products in food supplements and the nutraceuticals industry could provide a way of valorization in the transition to becoming more sustainable. In this context, the present study describes the phenolic profiling of sesame (*Sesamum indicum* L.) cake using reversed-phase high-performance liquid chromatography coupled to diode array detection and quadrupole-time-of-flight-mass spectrometry. Compounds were characterized based on their retention time, UV spectra, accurate mass spectrometry (MS) and MS/MS data along with comparison with standards, whenever possible, and the relevant literature. The characterized compounds (112 metabolites) belong to several classes, namely, phenolic acids (hydroxybenzoic acids and hydroxycinnamic acids), flavonoids, and lignans. Moreover, organic acids and some nitrogenous compounds were characterized. The total phenol content and the antioxidant activity of the cake extract were determined. This study provides useful information for the valorization of by-products from the sesame oil industry.

Keywords: *Sesamum indicum* L.; sesame cake; RP-HPLC–DAD–QTOF-MS; phenolic acids; lignans; flavonoids; agri-food residues

---

1. Introduction

Pedaliaceae is considered a small family, with 14 genera and 70 species. It is natively distributed in the Old World and commonly known as the sesame family [1]. Sesame (*Sesamum indicum* L.) is a prominent oil crop that is cultivated all over the world. It is believed that sesame originates from India [2]. Nevertheless, it has been present in the Ancient Egyptian civilization since the third century BC, where the ancient Egyptian used it for soothing asthma [3].

Sesame seeds are considered a rich source of proteins, dietary fibers, carbohydrates, fats, and vitamins [2,4]. Several studies investigated the phytochemical composition of sesame seeds and/or oil with a focus on lignans. For instance, Dachtler and coworkers applied online liquid chromatography coupled to nuclear magnetic resonance and mass spectrometry (MS) to characterize lignans in sesame oil, whereas Grougnet and others isolated lignans from sesame seeds and perisperms, respectively [5,6]. Regarding other phytochemicals, Hassan studied the physical characteristics of two Egyptian cultivars of sesame as well as their content of phenolic acids while taking into account the effect of roasting of seeds. Botelho et al. applied supercritical fluid extraction to black sesame seeds. The extracts showed significant amounts of unsaturated fatty acids and phytosterols, which were subjected for neuroprotective studies [7]. Moreover, sesame seeds showed several biological activities *viz.* antioxidant, hypolipidemic, hypocholesterolemic, antidiabetic, anticancer, antihypertensive, and cardioprotective activities [2,8,9]. Thus, more studies are required to establish the chemical composition of sesame since, as far as we know, most of the published studies on sesame focused on a short list of compounds.

Today, new efforts are addressed to valorize agri-industrial by-products in different ways. This includes obtaining extracts rich in functional secondary metabolites, which can be useful in the sustainable development of functional food supplements and nutraceuticals [10]. In accordance with the Food and Agriculture Organization (FAO) of the United Nations statistics, the global production of sesame is higher than 7 million tons, and around 2 million tons of sesame oil are produced. Consequently, there is a growing interest in the utilization of the cake by-product, which represents nearly 70% of the total production of the seeds [11]. The cake could maintain a large part of the phytochemical composition of the seeds and, thus, their functionality.

Therefore, the objective of this study was to perform an untargeted metabolic profiling of sesame cake, mainly focused on the structural elucidation of phenolic compounds, through reversed-phase (RP) high-performance liquid chromatography (HPLC) coupled to diode array detection (DAD) and quadrupole-time-of-flight (QTOF)-MS. In addition, the antioxidant activity *via* measurement of the total phenol content (TPC) and the Trolox equivalent antioxidant capacity (TEAC) was assayed. These results give new insights into the phenolic composition of this undervalued sesame by-product.

2. Materials and Methods

2.1. Chemicals and Reagents

Methanol, *n*-hexane, acetone, acetonitrile, and glacial acetic acid were purchased from Fisher Chemicals (Thermo Fisher Scientific, Waltham, MA, USA). The solvents used for extraction and characterization were of analytical and MS grade, respectively. Ultrapure water was obtained with a Milli-Q system (Millipore, Bedford, MA, USA). Folin & Ciocalteu's phenol reagent, sodium carbonate, ABTS (2,2′-azinobis (3-ethylbenzothiazoline-6-sulfonate)), Trolox (6-hydroxy-2,5,7,8-tetramethylchroman-2-carboxylic acid), potassium persulfate, L-tyrosine, and phenolic standards were purchased from Sigma-Aldrich (St. Louis, MO, USA), while L-tryptophan and L-phenylalanine were from Acros Organics (Morris Plains, NJ, USA). Kaempferide was purchased from Extrasynthese (Genay, France). The degree of purity of the standards was around 95% (*w*/*w*).

2.2. Samples Procurement and Extraction Procedures

The Egyptian sesame cultivar 'Giza 32' was kindly provided and identified by Agriculture Engineer Nadia Abdel-Azim, Egyptian Ministry of Agriculture and Land Reclamation (Giza, Egypt). Prior to the extraction, the seeds were ground (particle size around 1 mm) with Ultra Centrifugal Mill ZM 200, Retsch (Haan, Germany).

The first step of the extraction method was performed according to previous studies [12,13] with some modifications. For each one, 1 g of sample was firstly homogenized with 10 mL *n*-hexane and subjected to magnetic stirrer Agimatic-N (Jp Selecta, Barcelona, Spain) at room temperature for 30 min for defatting. Then defatted sesame cake was homogenized with 25 mL methanol/water (50:50, *v*/*v*) using an Ultra-TurraxIka T18 basic (Ika-Werke GmbH & Co. KG, Staufen, Germany), sonicated for 10 min at room temperature in an ultrasonic bath B3510 (40 kHz) (Branson, Danbury, CT, USA), and homogenized in the aforementioned magnetic stirrer at room temperature for 60 min. The mixture was finally centrifuged at 7155× *g* (8000 rpm) for 15 min and 5 °C using Sorvall ST 16 (Thermo Sci., ThermoFisher, Waltham, MA, USA) and the supernatant was collected. The pellet was re-extracted with 25 mL acetone/water (70:30, *v/v*). Both supernatants were combined and evaporated under vacuum using a rotary evaporator at 38 °C (Rotavapor R-200, BüchiLabortechnik, AG, Switzerland). Three independent extractions were performed. Finally, the dry cake extracts were dissolved in methanol/water (80:20, *v/v*), filtered (0.45 µm syringe filter, regenerated cellulose) and stored at −20 °C until analysis.

2.3. Analysis by RP-HPLC–DAD–ESI–QTOF-MS and -MS/MS

Analyses were made with an Agilent 1200 series rapid resolution (Santa Clara, CA, USA) equipped with a binary pump, an autosampler, and a diode array detector (DAD). Separation was carried out with a core-shell Halo C18 analytical column (150 mm × 4.6 mm, 2.7 µm particle size). The system was coupled to a 6540 Agilent Ultra-High-Definition (UHD) Accurate-Mass Q-TOF LC/MS equipped with an Agilent Dual Jet Stream electrospray ionization (Dual AJS ESI) interface.

Gradient elution was conducted with two mobile phases consisting of acidified water (0.5% acetic acid, *v/v*) (phase A) and acetonitrile (phase B) with a constant flow rate of 0.5 mL/min according to [13]. The gradient program was 0 min, 99% A and 1% B; 5.50 min, 93% A and 7% B; 11 min, 86% A and 14% B; 17.5 min, 76% A and 24% B; 22.50 min, 60% A and 40% B; 27.50 min, 0% A and 100% B; 28.5 min 0% A and 100% B; 29.5 min, initial conditions, which were finally maintained for 5.50 min for column equilibration (total run 35 min). The injection volume was 15 µL and three analyses were performed.

The operating conditions briefly were drying nitrogen gas temperature, 325 °C with a flow of 10 L/min; nebulizer pressure, 20 psig; sheath gas temperature, 400 °C with a flow of 12 L/min; capillary voltage, 4000 V; nozzle voltage, 500 V; fragmentor voltage, 130 V; skimmer voltage, 45 V; octapole radiofrequency voltage, 750 V. Data acquisition (2.5 Hz) in profile mode was governed *via* MassHunter Workstation software (Agilent technologies). The spectra were acquired in negative-ion mode over a mass-to-charge (*m/z*) range from 70 to 1500. The detection window was set to 100 ppm. Reference mass correction on each sample was performed with a continuous infusion of Agilent TOF biopolymer analysis mixture containing trifluoroacetic acid ammonium salt (*m/z* 112.9856 corresponding to the trifluoroacetic acid ion) and hexakis (1H, 1H, 3H-tetrafluoropropoxy) phosphazine (*m/z* 1033.9881 corresponding to the trifluoroacetic acid ammonium salt adduct ion). MS/MS experiments were performed in automatic mode, using the following collision energy values: *m/z* 100, 40 eV; *m/z* 500, 45 eV; *m/z* 1000, 50 eV; and *m/z* 1500, 55 eV.

MassHunter Qualitative Analysis B.06.00 (Agilent technologies) was used for data analysis following the strategy proposed by [10,12,14]. The characterization of compounds was performed by generation of candidate formula with a mass accuracy limit of 5 ppm. The MS score related to the contribution to mass accuracy, isotope abundance and isotope spacing for the generated molecular formula was set at ≥90. After the generation of the molecular formula, retention time (Rt), UV, and MS/MS spectra were also considered and compared with literature. Consequently, the following chemical structure databases were consulted: ChemSpider (http://www.chemspider.com), SciFinder Scholar (https://scifinder.cas.org), Reaxys (http://www.reaxys.com), PubChem (http://pubchem.ncbi. nlm.nih.gov), KNApSAcK Core System database (http://www.knapsackfamily.com/knapsack_jsp/ top.html), METLIN Metabolite Database (http://metlin.scripps.edu), Phenol-Explorer (www.phenol-explorer.eu), Dictionary of Natural Products (http://dnp.chemnetbase.com), Phytochemical dictionary of natural products database [15] and tracing available literature *via* Egyptian Knowledge Bank (https://www.ekb.eg/). Confirmation was made through a comparison with standards, whenever these were available in-house. Moreover, the peak area obtained by MS for each compound was determined to estimate the abundance of each characterized phenolic compound.

2.4. Total Phenol Content (TPC) Assay

The TPC of the extracts was determined in triplicate colorimetrically by Folin–Ciocalteu reagent [16] and modified according to [17] in 96-well polystyrene microplates (Thermo Fisher) and using a Synergy MxMonochromator-Based Multi-Mode Microplate reader (Bio-Tek Instruments Inc., Winooski, VT, USA). The absorbance of the solution was measured at a wavelength of 760 nm after incubation for 2 h in the dark and compared with a calibration curve of serially diluted gallic acid (GA). The results were expressed as GA equivalents (GAE). Analyses were done in triplicate.

2.5. Trolox Equivalent Antioxidant Capacity Assay

TEAC absorbance measurements were performed using the aforementioned microplate reader and following the procedure described by [12]. Briefly, $ABTS^+$ was produced by reacting ABTS stock solution with 2.45 mM potassium persulfate (final concentration). The mixture was kept in dark at room temperature for 24 h and the solution diluted with water until reaching an absorbance value of 0.70 (±0.03) at 734 nm. Afterwards, 300 µL of this solution and 30 µL of the cake extract were mixed and measured. Absorbance reading was compared to a standard calibration curve of Trolox and the results expressed as equivalents of Trolox (TE). Caffeic acid was used as a positive control. Analyses were done in triplicate.

3. Results and Discussion

3.1. Metabolic Profiling of Sesame Cake by RP-HPLC–DAD–QTOF-MS and -MS/MS

The sesame cake was subjected to core–shell RP-HPLC–DAD–ESI–QTOF-MS and -MS/MS analysis in negative ionization mode, providing Rt, experimental *m/z*, generated molecular formulae, mass error in ppm, mass score, double bond equivalents (DBE), UV maxima, and tandem mass fragments. For clarification, Tables 1 and 2 illustrate the aforementioned parameters for phenolic compounds and non-phenolic compounds, respectively. Moreover, Tables S1 and S2 (supplementary material) also detail metabolites class/subclass, plant species, family, and previously reported literature for phenolic compounds and non-phenolic compounds, respectively. The total number of characterized metabolites was 112, including 92 metabolites that were reported for the first time in sesame, with 20 new proposed structures. Figure 1 shows the base peak chromatogram of the cake extract, showing its complexity.

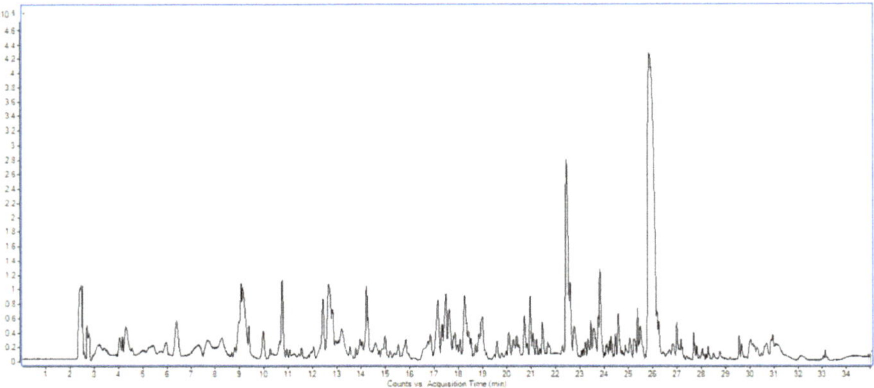

Figure 1. Base peak chromatogram of the cake of the Egyptian cultivar of sesame 'Giza 32'.

Table 1. Phenolic compounds characterized in the cake of the Egyptian cultivar of *Sesamum indicum* L. 'Giza 32'.

Number	Rt (min)	Experimental *m/z* [a] [M − H]⁻	Theoretical Mass (M)	Molecular Formula	Error (ppm)	Error (mDa)	Score	Main Fragments	DBE	UV (nm)	Proposed Compound [c]
1	7.33	169.0140	170.0215	$C_7H_6O_5$	2.09	0.36	84.05	N.D.	5	N.D.	Gallic acid *
2	8.74	137.0242	138.0316	$C_7H_6O_3$	0.51	0.07	96.91	N.D.	5	N.D.	Sesamol
3	9.02	329.0881	330.0951	$C_{14}H_{18}O_9$	−0.5	−0.2	92.03	167.0351, 123.0456	6	256sh, 286	Vanillic acid hexoside I
4	9.10	147.0454	148.0524	$C_9H_8O_2$	−1.2	−0.3	98.8	103.0556, 85.0290	6	255	Cinnamic acid
5	9.56	329.0878	330.0951	$C_{14}H_{18}O_9$	0.0	0.0	96.43	197.0457, 153.0193	6	N.D.	**Syringic acid pentoside**
6	9.91	343.1036	344.1107	$C_{15}H_{20}O_9$	−0.2	−0.1	99.74	163.0401, 119.0498	6	N.D.	**p-Coumaric acid hexoside I (in hydrated form)**
7	10.28	329.0879	330.0951	$C_{14}H_{18}O_9$	−1.0	−0.3	94.93	167.0349, 152.0117, 123.0450, 108.0218	6	257sh, 285	Vanillic acid hexoside II
8	10.66	329.0882	330.0951	$C_{14}H_{18}O_9$	−1.0	−0.3	97.69	167.0349, 152.0113, 123.0451, 108.0217	6	255sh, 285	Vanillic acid hexoside III
9	11.20	461.1302	462.1373	$C_{19}H_{26}O_{13}$	0.0	0.0	94.27	329.0876, 299.0775, 167.0345, 152.0194	7	N.D.	Vanillic acid pentosidehexoside
10	11.33	153.0194	154.0266	$C_7H_6O_4$	−0.64	−0.1	84.11	N.D.	5	257, 283sh	3,4-Dihydroxybenzoic acid * (protocatechuic acid)
11	11.83	359.0982	360.1057	$C_{15}H_{20}O_{10}$	0.4	0.2	99.54	197.0449, 182.0218, 166.9988, 153.0555, 138.0318, 123.0087	6	267	Syringic acid hexoside
12	12.35	325.0932	326.1002	$C_{15}H_{18}O_8$	−0.6	−0.2	99.12	163.0399, 119.0500	7	287	p-Coumaric acid hexoside II
13	12.72	325.0932	326.1002	$C_{15}H_{18}O_8$	−0.6	−0.2	99.12	163.0419, 119.0503	7	286	p-Coumaric acid hexoside III
14	13.58	487.1456	488.1530	$C_{21}H_{28}O_{13}$	−0.08	−0.04	80.23	341.1085, 179.0544	8	270	Caffeic acid hexosidedeoxyhexoside (cistanoside F)
15	13.93	355.1039	356.1107	$C_{16}H_{20}O_9$	−1.0	−0.4	99.18	193.0504, 178.0268, 134.0378, 119.0489	7	228, 288, 310	Ferulic acid hexoside isomer I
16	14.22	355.1038	356.1107	$C_{16}H_{20}O_9$	−0.7	−0.3	98.93	193.0507, 178.0269, 134.0374, 119.0513	7	228, 288, 310	Ferulic acid hexoside isomer II
17	14.30	153.0193	154.0266	$C_7H_6O_4$	0.33	0.05	85.78	N.D.	5	N.D.	2,5-Dihydroxybenzoic acid * (gentisic acid)
18	14.89	137.0242	138.0317	$C_7H_6O_3$	−0.1	0.0	93.21	N.D.	5	272	p-Hydroxybenzoic acid *
19	14.96	385.1136	386.1213	$C_{17}H_{22}O_{10}$	−0.5	−0.2	87.56	223.0611, 208.0373, 179.0706, 164.0472, 149.0238	7	290	Sinapic acid hexoside
20	14.99	353.0878	354.0951	$C_{16}H_{18}O_9$	0.7	0.3	99.38	N.D.	8	278	Caffeoylquinic acid * (chlorogenic acid)
21	15.12	531.1719	532.1792	$C_{23}H_{32}O_{14}$	0.0	0.1	96.82	385.1140, 223.0610, 179.0706, 165.0551, 150.0315	8	283	**Sinapic acid deoxyhexosidehexoside I**
22	15.16	517.1560	518.1635	$C_{22}H_{30}O_{14}$	0.1	0.0	95.85	193.0501, 175.0395, 149.0606, 134.0367	8	277	Ferulic acid dihexoside (sibiricose A5)
23	15.58	325.0930	326.1002	$C_{15}H_{18}O_8$	−0.30	−0.10	97.64	163.0398, 121.0293, 119.0498	7	228, 276	p-Coumaric acid hexoside IV
24	15.74	325.0930	326.1002	$C_{15}H_{18}O_8$	−0.38	−0.12	99.22	163.0391, 119.0498	7	282, 312	p-Coumaric acid hexoside V

Table 1. *Cont.*

Number	Rt (min)	Experimental m/z [a] $[M - H]^-$	Theoretical Mass (M)	Molecular Formula	Error (ppm)	Error (mDa)	Score	Main Fragments	DBE	UV (nm)	Proposed Compound [c]
25	16.61	325.0932	326.1002	$C_{15}H_{18}O_8$	−0.5	−0.2	97.58	193.0508, 134.0375	7	291	**Ferulic acid pentoside I**
26	16.75	167.0356	168.0426	$C_8H_8O_4$	−3.59	−0.60	94.34	N.D.	5	232, 264, 293	Vanillic acid *
27	16.81	355.1050	356.1107	$C_{16}H_{20}O_9$	−3.8	−1.4	93.17	193.0506, 178.0274, 134.0378, 119.0502	7	228, 288	Ferulic acid hexoside isomer III
28	16.88	593.1516	594.1585	$C_{27}H_{30}O_{15}$	−0.6	−0.4	98.31	503.1189, 473.1090, 383.0773, 353.0666	13	280, 323	LuteolinC-deoxyhexosideC-hexoside I
29	17.01	179.0349	180.0422	$C_9H_8O_4$	0.6	0.1	86.26	N.D.	6	233, 273	Caffeic acid *
30	17.15	593.1518	594.1585	$C_{27}H_{30}O_{15}$	−1.0	−0.6	98.52	533.1313, 503.1623, 473.1085, 383.0772, 353.0664	13	272, 330	LuteolinC-deoxyhexosideC-hexoside II
31	17.44	197.0453	198.0528	$C_9H_{10}O_5$	1.3	0.3	80.31	182.0218, 166.9984, 153.0182, 123.0088	5	273	Syringic acid *
32	17.51	593.1522	594.1585	$C_{27}H_{30}O_{15}$	−1.7	−1.0	97.71	533.1290, 503.1186, 473.1084, 383.0770, 353.0663	13	272, 330	LuteolinC-deoxyhexosideC-hexoside III
33	17.62	531.1724	532.1792	$C_{23}H_{32}O_{14}$	−0.8	−0.4	98.13	208.0329, 165.0560, 150.0323	8	278	**Sinapic acid deoxyhexosidehexoside II**
34	17.73	289.0718	290.0790	$C_{15}H_{14}O_6$	−0.33	−0.10	85.47	N.D.	9	234, 272	(-)-Epicatechin *
35	17.83	531.1733	532.1792	$C_{23}H_{32}O_{14}$	−2.1	−1.1	96.91	165.0558, 150.0322	8	277	**Sinapic acid deoxyhexosidehexoside III**
36	17.96	681.2400	682.2473	$C_{32}H_{42}O_{16}$	0.00	0.00	99.86	519.1854, 357.1337, 342.1101, 327.1238, 151.0393, 136.0156	12	280	Pinoresinol dihexoside I
37	18.24	563.1415	564.1479	$C_{26}H_{28}O_{14}$	−1.65	−0.93	96.8	545.1296, 503.1192, 473.1084, 443.0983, 413.0873, 383.0771, 353.0666, 117.0336	13	272, 327	Apigenin C-pentoside-C-hexoside I
38	18.30	121.0296	122.0368	$C_7H_6O_2$	−1.21	−0.15	99.65	92.0266	5.5	272	Benzoic acid
39	18.67	563.1413	564.1479	$C_{26}H_{28}O_{14}$	−1.01	−0.57	98.77	545.1297, 503.1188, 473.1086, 443.0983, 413.0876, 383.0766, 353.0661, 117.0351	13	272, 326	Apigenin C-pentoside-C-hexoside II
40	18.83	563.1414	564.1479	$C_{26}H_{28}O_{14}$	−1.19	−0.67	98.47	545.1292, 503.1186, 473.1081, 443.0976, 413.0874, 383.0766, 353.0661, 117.0337	13	269, 337	Apigenin C-pentoside-C-hexoside III
41	18.90	447.0938	448.1006	$C_{21}H_{20}O_{11}$	−2.10	−0.94	96.95	357.0618, 327.0510, 297.0405, 285.0407	12	272, 332	Luteolin C-hexoside I
42	18.93	563.1412	564.1479	$C_{26}H_{28}O_{14}$	−0.97	−0.55	98.82	545.1296, 503.1186, 473.1082, 443.0976, 413.0874, 383.0768, 353.0664, 117.0346	13	272, 333	Apigenin C-pentoside-C-hexoside IV

Table 1. *Cont.*

Number	Rt (min)	Experimental m/z [a] $[M - H]^-$	Theoretical Mass (M)	Molecular Formula	Error (ppm)	Error (mDa)	Score	Main Fragments	DBE	UV (nm)	Proposed Compound [c]
43	19.18	533.1303	534.1373	$C_{25}H_{26}O_{13}$	−0.45	−0.24	98.84	N.D.	13	274, 320sh	Apigenin di-C-pentoside I
44	19.44	563.1411	564.1479	$C_{26}H_{28}O_{14}$	−1.08	−0.61	98.75	545.1300, 503.1206, 473.1087, 443.0980, 413.0875, 383.0768, 353.0668	13	N.D.	Apigenin C-pentoside-C-hexoside V
45	19.48	325.0930	326.1002	$C_{15}H_{18}O_8$	−0.1	0.0	99.16	193.0506, 175.0973	7	N.D.	**Ferulic acid pentoside II**
46	19.60	447.0945	448.1006	$C_{21}H_{20}O_{11}$	−2.35	−1.05	96.73	357.0611, 327.0507, 297.0404, 285.0400	12	265, 334	Luteolin C-hexoside II
47	19.85	767.2402	768.2477	$C_{35}H_{44}O_{19}$	0.07	0.06	91.82	723.2460, 357.1341, 342.1102, 327.1236, 151.0399, 136.0165	14	276	**Pinoresinol malonyl dihexoside I**
48	19.92	533.1304	534.1373	$C_{25}H_{26}O_{13}$	−1.49	−0.80	97.89	515.1191, 503.1164, 473.1088, 443.0983, 413.0870, 383.0101, 353.0667, 117.0340	13	277	Apigenin di-C-pentoside II
49	19.95	767.2405	768.2477	$C_{35}H_{44}O_{19}$	−0.20	−0.15	97.21	723.2656, 681.2187, 561.1957, 357.1346, 342.1104, 327.1236, 151.0399, 136.0165	14	279	**Pinoresinol malonyl dihexoside II**
50	20.33	163.0399	164.0473	$C_9H_8O_3$	−0.14	−0.02	96.2	119.0501	6	284	p-Coumaric acid *
51	20.52	609.1463	610.1534	$C_{27}H_{30}O_{16}$	−0.32	−0.19	96.93	301.0825, 178.9975, 151.0029	13	N.D.	Quercetin 3-O-rutinoside (rutin) *
52	21.06	681.2399	682.2473	$C_{32}H_{42}O_{16}$	0.50	0.34	97.7	519.1519, 357.1325, 327.0853, 339.0883, 309.0763, 151.0387, 137.0249	12	280	Pinoresinol dihexoside II
53	21.13	463.0884	464.0955	$C_{21}H_{20}O_{12}$	−0.54	−0.25	96.41	N.D.	12	N.D.	Quercetin 3-O-β-ᴅ-glucopyranoside *
54	21.23	623.1994	624.2054	$C_{29}H_{36}O_{15}$	−1.64	−1.02	97.45	461.1689, 315.1084, 297.0981, 179.0351, 161.0240, 153.0567, 135.0452, 113.0244	12	280, 308	Verbascoside
55	21.27	639.1993	640.2003	$C_{29}H_{36}O_{16}$	−0.03	−0.02	96.46	477.1427, 331.0809, 297.0764, 179.0339, 161.0453	12	282	β-Hydroxyverbascoside (campneoside II)
56	21.30	463.0884	464.0955	$C_{21}H_{20}O_{12}$	−1.11	−0.52	97.47	N.D.	12	N.D.	Quercetin 3-O-β-D-galactopyranoside*
57	21.36	447.0929	448.1006	$C_{21}H_{20}O_{11}$	0.77	0.34	81.14	N.D.	12	N.D.	Luteolin 7-O-β-D-glucopyranoside *
58	21.48	223.0611	224.0685	$C_{11}H_{12}O_5$	0.90	0.20	83.94	N.D.	6	N.D.	Sinapic acid *
59	21.54	193.0506	194.0579	$C_{10}H_{10}O_4$	−0.4	−0.1	98.4	178.0271, 134.0371, 119.0502	7	230, 295sh, 322	Ferulic acid *
60	22.09	161.0245	162.0317	$C_9H_6O_3$	−1.2	−0.2	84.29	N.D.	7	N.D.	7-Hydroxycoumarin * (umbelliferone)

Table 1. *Cont.*

Number	Rt (min)	Experimental m/z [a] $[M - H]^-$	Theoretical Mass (M)	Molecular Formula	Error (ppm)	Error (mDa)	Score	Main Fragments	DBE	UV (nm)	Proposed Compound [c]
61	22.29	1017.3112	1018.3165	$C_{44}H_{58}O_{27}$	−1.44	−1.46	97.58	855.2379, 693.1894, 369.0920, 323.0935, 221.0642, 219.0625, 179.0543, 161.0422, 149.0425, 143.0341	16	286	**Sesaminol tetrahexoside I**
62	22.29	623.1981	624.2054	$C_{29}H_{36}O_{15}$	0.37	0.23	98.18	461.1659, 315.1088, 297.0969, 179.0354, 161.0241, 153.0546, 135.0453, 113.0243	12	280, 308	Isoverbascoside
63	22.34	681.2389	682.2473	$C_{32}H_{42}O_{16}$	2.13	1.45	94.88	519.1850, 357.1336, 327.0850, 297.0766, 161.0456, 151.0394, 149.0429, 137.0609	12	288	Pinoresinoldihexoside III
64	22.40	1017.3097	1018.3165	$C_{44}H_{58}O_{27}$	−0.19	−0.19	99.14	855.2565, 693.2027, 369.0976, 323.0934, 221.0661, 219.0625, 179.0559, 161.0451, 149.0456	16	284	**Sesaminol tetrahexoside II**
65	22.56	841.2785	842.2857	$C_{38}H_{50}O_{21}$	−1.42	−1.20	97.87	679.2245, 517.1755, 485.1514, 355.1189, 323.0984, 221.0668, 179.0563, 161.0459, 149.0455, 121.0295, 89.0244	14	285	**Xanthoxylol trihexoside**
66	22.93	163.0399	164.0473	$C_9H_8O_3$	1.64	1.27	86.8	N.D.	6	284	*m*-Coumaric acid *
67	23.15	927.2789	928.2849	$C_{41}H_{52}O_{24}$	−1.26	−1.17	97.47	883.2852, 841.2758, 823.2653, 679.2213, 661.2124, 485.1489, 467.1399, 355.1182, 323.0980, 221.0659, 161.0457, 179.0552, 177.0921, 149.0448, 121.0293	16	288	**Xanthoxylol malonyl trihexoside**
68	23.26	447.0942	448.1006	$C_{21}H_{20}O_{11}$	−2.16	−0.97	85.48	N.D.	12	N.D.	Quercetin 3-*O*-rhamnopyranoside *
69	23.32	855.2585	856.2637	$C_{38}H_{48}O_{22}$	−2.0	−1.7	96.4	693.2013, 485.1485, 369.0965, 323.0980, 221.0663, 219.0668, 179.0558, 161.0455, 149.0453, 143.0347, 119.0350	15	283	Sesaminol trihexoside I
70	23.45	855.2578	856.2637	$C_{38}H_{48}O_{22}$	−1.8	−1.5	96.86	693.2329, 531.1494, 485.1508, 369.0978, 221.0664, 219.0657, 179.0558, 161.0457, 149.0456, 143.0349, 119.0349	15	290	Sesaminol trihexoside II

Table 1. *Cont.*

Number	Rt (min)	Experimental m/z [a] $[M - H]^-$	Theoretical Mass (M)	Molecular Formula	Error (ppm)	Error (mDa)	Score	Main Fragments	DBE	UV (nm)	Proposed Compound [c]
71	24.25	871.2515	872.2586	$C_{38}H_{48}O_{23}$	0.4	0.3	98.91	709.1974, 691.1869, 529.1376, 485.1509, 385.0926, 323.0985, 221.0665, 179.0556, 165.0192, 161.0457, 143.0347, 149.0452, 137.0243, 119.0349, 89.0243	15	292	**Hydroxysesamolin trihexoside**
72	24.32	531.1513	532.1581	$C_{26}H_{28}O_{12}$	−0.7	−0.4	99.00	337.0929, 323.0776, 193.0503, 178.0267, 175.0398, 149.0607, 134.0370	13	292	**Diferuloyl hexoside**
73	24.45	317.0305	318.0388	$C_{15}H_{10}O_8$	−3.7	−1.2	82.25	N.D.	11	N.D.	Myricetin *
74	26.45	285.0408	286.0477	$C_{15}H_{10}O_6$	−1.2	−0.3	99.42	N.D.	11	288, 340	Luteolin *
75	26.65	301.0349	302.0420	$C_{15}H_{10}O_7$	2.1	0.6	90.98	N.D.	11	N.D.	Quercetin *
76	27.01	593.1884 [b]	594.1949	$C_{28}H_{34}O_{14}$	−1.3	−0.8	98.05	371.1143, 356.0911, 233.0817, 161.0457, 138.0323	12	283	Sesamolinolhexoside
77	27.19	633.1829	634.1898	$C_{30}H_{34}O_{15}$	−0.3	−0.2	99.43	N.D.	14	286	**Sesaminol dipentoside I**
78	27.43	633.1824	634.1898	$C_{30}H_{34}O_{15}$	0.2	0.1	99.47	501.1317, 369.0969, 339.0969, 219.0646, 135.0341	14	285	**Sesaminol dipentoside II**
79	27.67	635.1984	636.2054	$C_{30}H_{36}O_{15}$	−0.4	−0.2	98.82	371.1132	13	285	**Sesamolinol dipentoside I**
80	27.82	635.1991	636.2054	$C_{30}H_{36}O_{15}$	−1.1	−0.7	96.65	371.1128	13	286	**Sesamolinol dipentoside II**
81	27.85	269.0457	270.0528	$C_{15}H_{10}O_5$	−1.0	−0.3	83.23	N.D.	11	N.D.	Apigenin *
82	27.94	635.1980	636.2054	$C_{30}H_{36}O_{15}$	0.2	0.1	99.43	371.1135	13	286	**Sesamolinol dipentoside III**
83	28.19	285.0403	286.0477	$C_{15}H_{10}O_6$	0.7	0.0	84.77	N.D.	11	N.D.	Kaempferol *
84	28.32	271.0610	272.0685	$C_{15}H_{12}O_5$	0.7	0.2	99.86	N.D.	10	N.D.	Naringenin *
85	29.81	255.0664	256.0736	$C_{15}H_{12}O_4$	−0.23	−0.06	99.3	213.0551, 171.0440, 151.0026, 107.0136, 103.0534, 83.0134	10	288	Pinocembrin
86	29.84	299.0566	300.0634	$C_{16}H_{12}O_6$	−0.6	−0.2	89.67	N.D.	11	N.D.	Kaempferide *

[a] Detected ions were $[M-H]^-$. [b] Detected ion was acetic acid adduct $[M+CH_3COOH - H]^-$. [c] Isomers are denoted with letter codes I, II, etc. * Identification confirmed by comparison with standards; N.D., below 5 mAU or masked by compound with higher signal. Compounds in bold letter indicate new proposed structures.

Table 2. Non-phenolic compounds characterized in the cake of the Egyptian cultivar of *Sesamum indicum* L. 'Giza 32'.

Number	tR (min)	Experimental m/z^a [M − H]$^-$	Theoretical Mass (M)	Molecular Formula	Error (ppm)	Error (mDa)	Score	Main Fragments	DBE	UV (nm)	Proposed Compound [b]
1′	2.57	195.0513	196.0583	$C_6H_{12}O_7$	−1.3	−0.3	99.39	135.0305	1	N.D.	Gluconic/Galactonic acid I
2′	2.59	131.0463	132.0535	$C_4H_8N_2O_3$	0.11	0.01	97.72	114.0110, 113.0358	2	N.D.	Asparagine
3′	2.65	195.0513	196.0583	$C_6H_{12}O_7$	−0.9	−0.2	99.19	135.0297	1	N.D.	Gluconic/Galactonicacid II
4′	2.65	665.2144	666.2219	$C_{24}H_{42}O_{21}$	0.54	0.36	98.67	503.1618, 341.1086, 179.0558	4	N.D.	Sesamose
5′	2.96	191.0198	192.027	$C_6H_8O_7$	−0.2	−0	99.31	173.0087, 111.0089	3	N.D.	Citric acid I
6′	3.15	133.0143	134.0215	$C_4H_6O_5$	−0.7	−0.1	99.61	115.0037	2	N.D.	Malic acid I
7′	3.45	133.0141	134.0215	$C_4H_6O_5$	0.89	0.27	99.34	115.0039	2	N.D.	Malic acid II
8′	3.96	191.0197	192.027	$C_6H_8O_7$	0.17	0.03	99.75	173.0086, 111.0088	3	N.D.	Citric acid II
9′	4.02	147.0296	148.0372	$C_5H_8O_5$	1.3	0.19	97.42	103.0400	2	N.D.	Citramalic acid
10′	4.08	129.0191	130.0268	$C_5H_6O_4$	1.9	0.25	99.4	85.0297	3	N.D.	Itaconic acid
11′	4.27	191.0203	192.027	$C_6H_8O_7$	0.21	0.04	99.93	111.0086	3	N.D.	Citric acid III
12′	4.52	130.0872	131.0949	$C_6H_{13}NO_2$	1.49	0.2	99.47	112.986	1	N.D.	Leucine/Isoleucine
13′	5.33	180.0663	181.0745	$C_9H_{11}NO_3$	−1.7	−0.3	94.97	163.0406	5	264	Tyrosine *
14′	5.51	130.0874	131.0949	$C_6H_{13}NO_2$	−0.4	−0.1	98.93	112.9860	1	N.D.	Leucine/Isoleucine
15′	6.29	611.1454	612.152	$C_{20}H_{32}N_6O_{12}S_2$	−0.9	−0.6	98.87	481.1002, 338.0483, 306.0761, 288.0658, 254.0780, 179.0461, 128.0353	8	N.D.	Oxidized glutathione (Glutathione disulfide)
16′	7.00	171.0303	172.0372	$C_7H_8O_5$	−3.4	−0.6	95.58	127.0403	4	230	(−)-3-Dehydroshikimic acid
17′	7.67	191.0569	192.0634	$C_7H_{12}O_6$	−3.6	−0.7	96.92	147.0663, 129.0557, 101.0610	2	N.D.	Quinic acid I
18′	8.19	191.056	192.0634	$C_7H_{12}O_6$	0.1	0.0	99.64	147.0655, 129.0551, 101.0604	2	N.D.	Quinic acid II
19′	9.12	164.0718	165.0790	$C_9H_{11}NO_2$	−0.4	−0.1	99.53	147.0455, 129.0557, 85.0297	5	255	Phenylalanine *
20′	10.16	218.1034	219.1107	$C_9H_{17}NO_5$	0.1	0.0	98.6	146.0819	2	N.D.	Pantothenic acid (Vit B5) I
21′	10.53	218.1038	219.1107	$C_9H_{17}NO_5$	−1.4	−0.3	98.6	146.0822	2	N.D.	Pantothenic acid (Vit B5) II
22′	11.39	382.1003	383.1077	$C_{14}H_{17}N_5O_8$	−0.1	−0.1	98.1	266.0892, 250.574, 206.0679, 134.0468, 115.0034	9	265	Succinyladenosine
23′	12.72	529.1834	530.19	$C_{26}H_{30}N_2O_{10}$	−0.9	−0.5	99.2	203.0820, 159.0924, 142.0655, 116.0500	13	279, 287sh	Tryptophan derivative
24′	14.30	175.0611	176.0685	$C_7H_{12}O_5$	0.62	0.11	99.62	115.04	2	N.D.	Isopropylmalic acid I
25′	14.61	175.061	176.0685	$C_7H_{12}O_5$	1.41	0.25	99.21	115.04	2	N.D.	Isopropylmalic acid II
26′	24.03	187.0979	188.1049	$C_9H_{16}O_4$	−1.8	−0.3	98.65	125.097	2	N.D.	Azelaic acid

[a] Detected ions were [M − H]$^-$. [b] Isomers are denoted with letter codes I, II, etc. * Identification confirmed by comparison with standards; N.D., below 5 mAU or masked by compound with higher signal.

3.1.1. Hydroxybenzoic Acids

Qualitatively, phenolic acids were the most abundant phenolic compounds by 40 metabolites. They are divided into hydroxybenzoic acids (13) and hydroxycinnamic acids (27).

Concerning hydroxybenzoic acids, they can be classified into non-hydroxylated (benzoic acid), mono-hydroxylated (*p*-hydroxybenzoic acid), di-hydroxylated (gentisic, protocatechuic, and vanillic acid derivatives), and tri-hydroxylated (gallic acid and syringic acid derivatives). It bears noting that five of them were confirmed with standards. In brief, *m/z* 121.03 exerted a neutral loss of CO and UV absorbance at λ_{max} 272 nm. It was characterized as benzoic acid and has been previously reported in sesame [4]. The compound with a *m/z* value of 137.02 was confirmed to be *p*-hydroxybenzoic acid upon comparison with a standard. Similarly, two compounds showed a molecular formula of $C_7H_6O_4$ and *m/z* of 153.02. By comparing with standards, they were confirmed as protocatechuic acid (3,4-dihydroxybenzoic acid) and gentisic acid (2,5-dihydroxybenzoic acid).

In regard to *O*-methylated derivatives of dihydroxybenzoic acids, five derivatives of vanillic acid were characterized. Vanillic acid was observed with *m/z* 167.04, and was confirmed with standard. It has been reported before in sesame [18]. In addition, three isomers of vanillic acid hexoside ($C_{14}H_{18}O_9$) were observed at Rt 9.02, 10.28, and 10.66 min. They revealed the neutral loss of the hexose moiety (162 Da) and CO_2 (44 Da), complying with the typical decarboxylation of phenolic acids [10,12,17]. They were reported for the first time in *S. indicum*, nevertheless, they were reported in the genus *Sesamum* in accordance with the Dictionary of Natural Products database (Table S1). Vanillic acid pentoside hexoside was characterized with *m/z* 461.13 and sequential loss of pentose (*m/z* 329.09), hexose (*m/z* 167.03, i.e., vanillic acid ion), and methyl (*m/z* 153.02) moieties. It was detected for the first time in sesame (Table 1 and Table S1). Concerning tri-hydroxylated benzoic acids, both gallic and syringic acids were observed with *m/z* 169.01 and 197.05, respectively. These assignments were confirmed with standards. It is worth noting that the compound with an *m/z* value of 329.09 and molecular formula $C_{14}H_{18}O_9$ exerted the neutral loss of pentose with an aglycone fragment of (*m/z* 197.05) followed by fragmentation of syringic acid. It was tentatively identified as syringic acid pentoside and proposed as a new structure. Similarly, compound at *m/z* 359.10 ($C_{15}H_{20}O_{10}$) showed a neutral loss of a hexose moiety with syringic acid fragmentation pattern. In this way, two consecutive losses of CH_3 were observed until fragment 166.10, indicating methoxy substituents. Moreover, the loss of CO_2 from the carboxyl moiety was also observed (*m/z* 153.06 and 138.03) followed by losses of CH_3 (Figure 2a) [17]. Therefore, it was identified as syringic acid hexoside, which was described for the first time in sesame.

3.1.2. Hydroxycinnamic Acids

Regarding hydroxycinnamic acid, it was found in free form, conjugated with quinic acid or with sugars. The occurrence of *p*-coumaric acid, *m*-coumaric acid, chlorogenic acid (caffeoylquinic acid), caffeic acid, ferulic acid, and sinapic acid were unequivocally confirmed with standards enabling characterization validation, which was in agreement with previous studies [19]. It is worth noting that both *m*-coumaric acid and chlorogenic acid were described for the first time in sesame. Moreover, cinnamic acid (21, *m/z* 147.05) was observed with the neutral loss of CO_2 from the carboxylate moiety and water (Table 1 and Table S1).

Five isomers of *p*-coumaric acid hexosides were detected expressing the neutral loss of hexose moieties releasing aglycones of *m/z* 163.04 followed by decarboxylation (*m/z* 119.05), except for the hydrated form (M−H+H_2O) and, hence, dehydration occurred firstly [13]. In the same manner, three isomers of ferulic acid hexoside ($C_{16}H_{20}O_9$, *m/z* 355.10) and sinapic acid hexoside ($C_{17}H_{22}O_{10}$, *m/z* 385.11) were observed (Table 1 and Table S1). As an example, the fragmentation of the latter compound is shown in Figure 2b. Another compound was described as ferulic acid dihexoside (sibiricose A5). In addition, seven novel hydrocinnamic acids were tentatively identified as ferulic acid pentoside isomers (I–III), sinapic acid deoxyhexoside hexoside isomers (I–III), and diferuloyl hexoside (Table 1 and Table S1). The latter was characterized by the presence of fragment ions at *m/z* 337.0929 ($C_{16}H_{17}O_8^-$) and 193.0503 ($C_{10}H_9O_4^-$), i.e., feruloylhexosyl and ferulic acid, respectively. Ferulic acid ion showed the typical decarboxylation and demethylation of this type of compound in MS/MS (Figure 2c), as for sinapic acid. Remarkably, most of these compounds are described for the first time in sesame (Table S1).

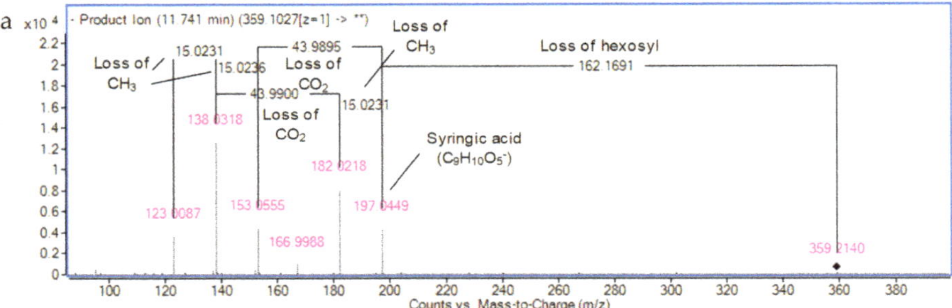

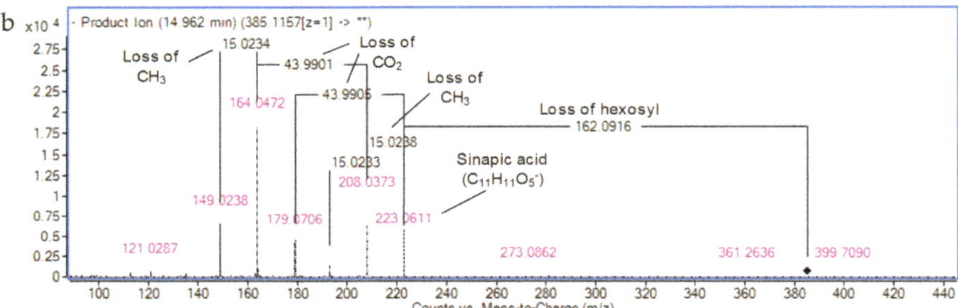

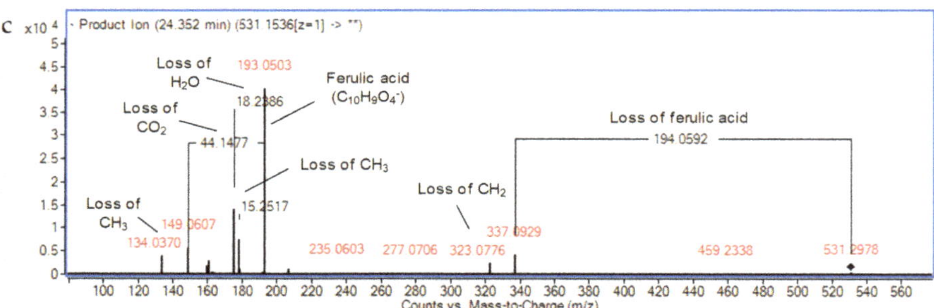

Figure 2. Fragmentation patterns of (**a**) syringic acid hexoside, (**b**) sinapic acid hexoside, and (**c**) diferuloyl hexoside.

Moreover, caffeoyl derivatives were found. Based on molecular formula ($C_{21}H_{28}O_{13}$) and fragmentation pattern with successive loss of deoxyhexose and hexose, the caffeic acid derivate cistanoside F ($C_{21}H_{28}O_{13}$) was characterized. Furthermore, three caffeoylphenylethanoid derivatives were observed. Verbascoside and isoverbascoside expressed the neutral loss of caffeoyl moieties and deoxyhexoses through the presence of the ion m/z 461.17 and m/z 315.11, respectively. β-Hydroxyverbascoside was identified with similar fragmentation pattern as of verbascoside with additional loss of hydroxyl group (16 Da) [19]. These caffeoyl derivatives were isolated before from sesame [20,21].

3.1.3. Lignans

Lignans are natural phenolic compounds that possess several biological activities, especially antioxidant and estrogenic activities. A simple lignan is composed of two phenyl propanoid derivatives (C6–C3) linked through a β–β-linkage [22]. In this context, 18 lignan derivatives were identified in cake of sesame, which belonged to the furofuran subclass of lignans. As a matter of fact, the in-depth analysis of tandem MS data made it possible to preliminary predict lignan structures, whose presence in sesame were not discovered and could yet contribute to its bioactivity (Table 1). For that, studies using tandem MS/MS were consulted for analog structures [22–25]. All of them were glycosides, and the loss of each sugar was observed until the product ions of the aglycones were released and hence observed [22].

Briefly, three isomers of pinoresinol dihexoside were detected showing subsequent losses of two hexosyl moieties with aglycone fragmentation showing the fragment m/z 151.04 ($C_8H_7O_3^-$) due to cleavage of the tetrahydrofuran ring followed by methyl loss from the guaiacyl moiety [22,24,25] (Table 1 and Table S1, Figure 3a). Similarly, two isomers, (m/z 767.24, $C_{35}H_{44}O_{19}$), resembled the aforementioned fragmentation with additional loss of a malonyl moiety (CO_2 and an acetyl moiety CH_2CO) (86 Da). Consequently, they were characterized as pinoresinol malonyl dihexoside (I–II). In the same manner, xanthoxylol trihexoside and xanthoxylol malonyl trihexoside were tentatively characterized with the observation of the loss of CH_3OH (32 Da) (m/z 323.10). In addition, minor fragments were observed at m/z 149.05 from the aglycone corresponding to methylenedioxyphenyl-CO ($C_8H_5O_3^-$) [23] and the counterpart after the loss of CO (m/z 177.09). The ion m/z 121.03 ($C_7H_5O_2^-$, methylenedioxyphenyl) was also observed according to [22,23], as well more abundant ions from sugars such as m/z 179.06 (hexose, $C_6H_{12}O_6$), 161.05 (hexosyl, $C_6H_{10}O_5$), and 89.02 ($C_3H_5O_3^-$). For clarification, Figure 3b illustrates the fragmentation pattern of xanthoxylol malonyl trihexoside. Although these glycosides have been reported here for the first time, and aglycone was characterized by Fukuda et al. [25].

In regard to sesaminol derivatives, two isomers of sesaminol trihexoside (I–II) and sesaminol tetrahexoside (I–II) were identified with the observation of m/z 149.05 characterizing furofurano lignans [23], as commented upon before, as well as the counterpart at m/z 219.06 ($C_{12}H_{11}O_4^-$). Moreover, two isomers of sesaminol dipentoside were identified, also showing a fragment at m/z 135.03; i.e.,methylenedioxyphenyl–CH_2 ($C_8H_7O_2^-$) [23]. The acetic acid adduct of sesamolinol hexoside (m/z 593.19, $C_{28}H_{34}O_{14}$) and three new isomers of sesamolinol dipentoside were observed with the observation of the aglycone at m/z 371.11 (Table 1 and Table S1). In the first case, fragment ions at m/z 138.0323 ($C_7H_6O_3^-$) and m/z 233.0817 ($C_{13}H_{13}O_4^-$) were detected, corresponding to the fragmentation of the aglycone structure, while in the rest of cases, the fragmentation of the aglycone was poor under the MS/MS conditions used for the assay. Finally, hydroxysesamolin trihexoside was tentatively identified with the presence of aglycone at m/z 385.09. The aglycone part showed MS/MS product ions at 137.0244 ($C_7H_5O_3^-$), which could indicate a similarity to sesamolinol structure, but also product ions at 165.0192 ($C_8H_5O_4^-$) and 149.0452 ($C_8H_5O_3^-$), which could indicate the presence of a hydroxylated methylenedioxyphenyl–CO moiety.

3.1.4. Coumarins

Regarding coumarins, umbelliferone (7-hydroxycoumarin) was unequivocally confirmed with a standard.

3.1.5. Flavonoids

A total of 26 flavonoids were characterized in the sesame cake extract, being classified mainly into a flavan-3-ol, flavanones (2), flavones (15), and flavonols (8) (Table 1). It is worth noting that (−)-epicatechin, naringenin, luteolin 7-*O*-β-D-glucopyranoside, luteolin, apigenin, quercetin, rutin, quercetin 3-*O*-β-D-glucopyranoside, quercetin 3-*O*-β-D-galactopyranoside, quercetin 3-*O*-rhamnopyranoside, myricetin, kaempferol, and kaempferide were identified through comparison with standards. All of them were described for the first time in the genus *Sesamum*, except for apigenin and luteolin 7-*O*-β-D-glucopyranoside, according to the phytochemical dictionary database [15].

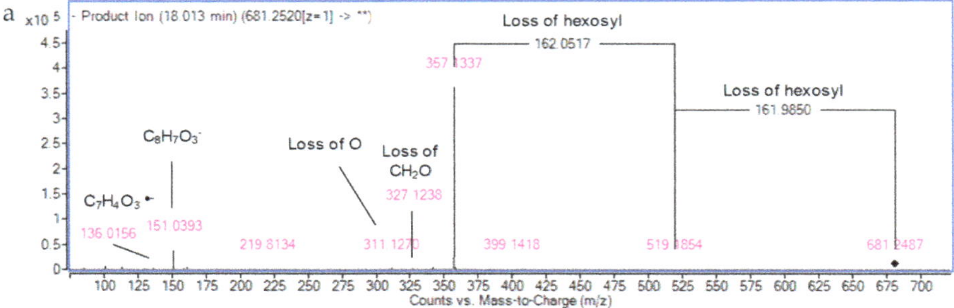

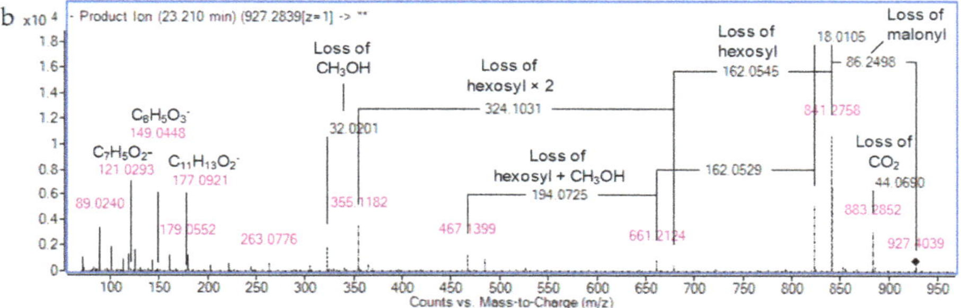

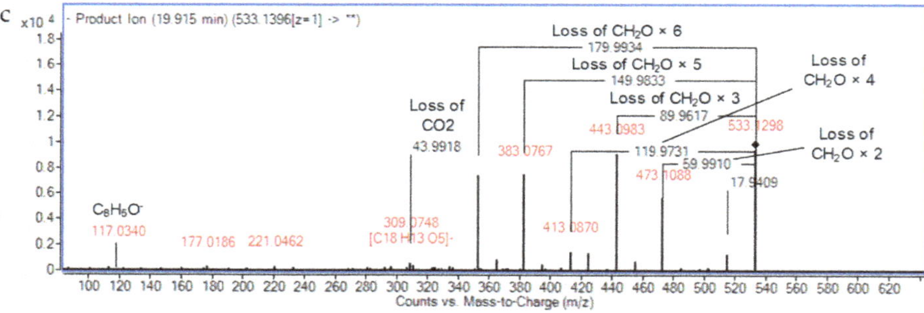

Figure 3. Patterns of (**a**) pinoresinol dihexoside I, (**b**) xanthoxylol malonyl trihexoside, and (**c**) apigenin di-*C*-pentoside II.

Additionally, the fragmentation pattern of compound at *m/z* 255.07 (Rt 29.81 min, $C_{15}H_{12}O_4$) revealed the common fragment ion released after retro Diels–Alder fission and retrocyclization at *m/z* 151.00 ($C_7H_3O_4$) ($^{1,3}A^-$) by this type of compound [26]. Moreover, the ion fragment with *m/z* at 103.05 ($C_8H_7^-$) could be derived from band B ($^{1,3}B^-$). Besides the fragmentation pattern, it showed UV absorbance at λ_{max} 288 nm, suggesting a flavanone nucleus [12]. Therefore, it was tentatively identified as pinocembrin, which was described for the first time in sesame.

The occurrence of *C*-glycosides of flavones was noticed with 12 derivatives of either luteolin or apigenin, which were observed for the first time in sesame. They were characterized by the presence of prominent fragment ions after the characteristic sequential loss of 90 ($C_3H_6O_3$) and/or 120 Da ($C_4H_8O_4$), in agreement with previous studies [12,14,17,19,27]. As an example, two isomers of luteolin *C*-hexoside were identified, exerting characteristic fragments at *m/z* 357.06 and 327.05, respectively. Similarly, three isomers of luteolin *C*-deoxyhexoside-*C*-hexoside were tentatively identified based on comparing their fragmentation pattern and UV absorbance with reported literature [27]. As for apigenin derivative, five isomers of apigenin *C*-pentoside-*C*-hexoside were observed, showing fragmentation patterns and UV absorbance of *C*-flavones as described in reported studies [15,17]. Moreover, a minor fragment ion at *m/z* 117.03 (C_8H_5O) ($^{1,3}B^-$) was observed, suggesting that the aglycone is apigenin. Similarly, two isomers of apigenin di-*C*-pentoside were tentatively characterized. As Figure 3c shows, the consequent neutral loss of sugar fragments (30–180 Da) was observed (Table 1 and Table S1, Figure 3c).

3.1.6. Others (Non-Phenolic Compounds)

A total of 17 organic acids were observed in the cake of the sesame, namely gluconic/galactonic acids, citric acid (I–III), malic acid (I–II), citramalic acid, itaconic acid, (−)-3-dehydroshikimic acid, quinic acid (I–II), pantothenic acid (I–II), isopropylmalic acid (I–II), and azelaic acid. Their fragmentation patterns were in agreement with reported studies [12,14,17,19,28–30] (Table 2 and Table S2). All of the identified organic acids are reported for the first time in sesame.

Regarding nitrogenous compounds, it is worth mentioning that five amino acids were characterized, *viz.* asparagine, leucine/isoleucine, tyrosine, and phenylalanine. Their fragmentation patterns were characterized by deamination and/or decarboxylation [13,15,17]. In addition, both tyrosine and phenylalanine were confirmed with standards. Furthermore, a peptide was observed (Rt 6.29 min, *m/z* 611.1454, $C_{20}H_{32}N_6O_{12}S_2$) exerting the loss of a glutathione moiety (*m/z* 306.08) followed by a loss of SH_2 from the cysteinyl group. Finally, the product ion of the glutamyl moiety was observed at *m/z* 128.04. It was compared with data on the METLIN database to be described as oxidized glutathione (GSSG), indicating the presence of reduced glutathione (GSH) in the cake of sesame, which is easily auto-oxidized to GSSG during sample preparation and/or analysis [31]. In fact, GSH is considered to be a powerful cellular antioxidant that prevents oxidative stress in biological systems and, hence, prevents the onset and progression of many serious diseases such as diabetes mellitus, cancer, and Alzheimer's disease [31]. Furthermore, a derivative of tryptophan (*m/z* 529.18, $C_{26}H_{30}N_2O_{10}$) was observed as well as succinyladenosine (*m/z* 382.10, $C_{14}H_{17}N_5O_8$), a nucleoside derivative. It bears noting that the characteristic tetrasaccharide sesamose could be the ion with *m/z* 665.21, presenting subsequent losses of hexosyl moieties in MS/MS (Table 2 and Table S2).

3.2. TPC, TEAC, and Phenolic Abundance

The extract of sesame cake showed a total phenol content of 1.9 ± 0.3 mg GAE/g cake extract. In fact, this value is even beyond results by Mohadaly et al. [18], where total phenol contents were assayed of single different solvents cake extracts of the Egyptian cultivar 'Shandweel-3'. The value of TPC ranged from 0.1 (petroleum ether extract) to 0.8 (methanol extract) mg GAE/g cake extract. This could be attributed to the combined solvent extraction accompanied with ultra-sonication, which enhances the extraction process [12,32]. In regard to the TEAC assay, the extract expressed a value of 2.65 ± 0.08 µmol TE/g of cake extract. In fact, Janu et al. [33] focused on the antioxidant activity of the sesame oil, which was found to be 0.004 µg TE/mL oil (i.e., around 0.02 µmol TE/g oil) indicating the value of the cake as an agri-industrial by-product that needs further attentions for its antioxidant potential as well as other biological activities.

To evaluate the contribution of phenolic compounds, a summary of the characterization results is shown in Figure 4. In the perspective of subclasses, flavonoids were the most abundant, representing 38.3% of the total characterized phenolic metabolites followed by hydroxycinnamic acids and then lignans (Figure 4a). Similarly, flavonoids and hydroxycinnamic acids were also the most representative families in qualitative terms (Figure 4b).

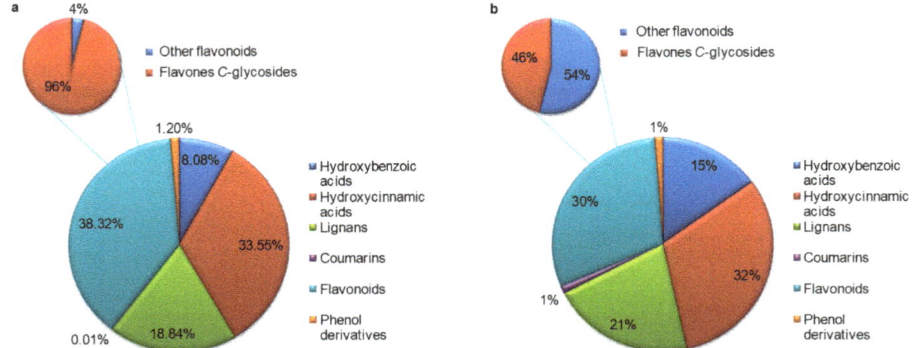

Figure 4. Summary of the characterization results on sesame phenolic compounds: (**a**) relative abundance (%) and (**b**) qualitative classification (%).

The exploration of alternative strategies for the treatment of many chronic diseases such as cancer, diabetes, and heart and liver diseases continues to attract scientists in discovering drugs derived from plant origins [34–36]. In fact, there is growing attention in the valorization of agri-food residues to provide new functional ingredients with bioactivities for sustainability of the agri-industry. It bears noting that such by-products represent around 40% of total plant foods [10,37]. For that, the elucidation of the potential bioactive phytochemicals is a requirement. In this regard, the application of UHPLC–QTOF-MS enabled us to characterize 86 phenolic compounds in sesame cake and, hence, as far as we know, this is the first study providing comprehensive phenolic profiling of sesame cake. In addition, the antioxidant activity of the sesame cake extract was determined by the TEAC method. In this regard, furofurano lignans possess anticancer, cardiovasculoprotective, neuroprotective, antioxidant, and anti-inflammatory activities. Moreover, they are metabolized by gut microflora into enterolactone and enterodiol, which are considered phytoestrogens [38]. Thus, these compounds could contribute to the antioxidant activity of the extract.

In this line, a previous study on sesame cake showed that the main contributors to the antioxidant activity were sesamol and water- (sesaminol tri- and di-glucoside) and lipid-soluble lignans (sesamin and sesamolin), but the extraction procedure was based on Soxhlet extraction with methanol [39]. In this sense, our results revealed the presence of sesamol and a wider range of lignan glycosides, but there were no free lignans, which could be due to the use of more polar extraction conditions and the removal of the fatty phase. Moreover, *C*-glycosides were the most abundant compounds both as a subclass, accounting for around 37.3% in relative abundance, and individually, i.e., apigenin *C*-pentoside-*C*-hexoside I (8.8%) followed by luteolin *C*-deoxyhexoside-*C*-hexoside III (7.0%), luteolin *C*-deoxyhexoside-*C*-hexoside II (6.7%), and apigenin *C*-pentoside-*C*-hexoside IV (5.1%). As a matter of fact, it seems that *C*-glycosylation enhances antioxidant capacity, where the hydroxyl group and metal chelation sites of flavones are free [14,40]. Thus, these compounds could be the highest contributors to the antioxidant activity of the extract, agreeing with Zhou et al. [41]. These authors highlighted that the antioxidant activity of sesame cake extracts was associated with the total content of flavonoids, but these compounds were not characterized. Furthermore, oxidized glutathione was detected for the first time in the cake of sesame, which could be produced from reduced glutathione during sample preparations. It is considered a strong marker for the antioxidant potential of this agri-industrial byproduct.

4. Conclusions

In this study, core–shell RP-HPLC–DAD–ESI–QTOF-MS and -MS/MS methods were employed to analyze the cake of the Egyptian cultivar of sesame 'Giza 32'. A total of 112 metabolites were characterized in sesame cake, and among them, 86 were phenolic compounds. The observed lignans were of furofurano type and, among them, 12 lignans are considered to be new proposed structures. Moreover, this is the first report showing the conjugation of malonyl moieties to lignans in Pedaliaceae. With regard to the characterized flavonoids, they were classified into flavones (15), flavonols (8), flavanones (2), and a flavan-3-ol. *C*-Glycosides of flavones have been reported here for the first time in sesame. This type of flavonoid was the most abundant. Furthermore, the antioxidant activity of the sesame cake extract was determined by TEAC method and, hence, our results suggest that not only sesamol and lignans, but also *C*-glycosides and other compounds could contribute to this bioactivity. Consequently, further studies are required for the development of food supplements and nutraceuticals from sesame cake to widen its applicability and to move into a more sustainable industry with zero waste.

Supplementary Materials: The following are available online at http://www.mdpi.com/2304-8158/8/10/432/s1, Table S1: Phenolic compounds characterized in the cake of the Egyptian cultivar of sesame 'Giza 32', Table S2: Non-phenolic compounds characterized in the cake of the Egyptian cultivar of sesame 'Giza 32'.

Author Contributions: Investigation, R.H.M. and M.d.M.C.; Methodology, M.d.M.C.; Data analysis, R.H.M.; Validation, R.H.M., M.d.M.C.; Writing—Original Draft, R.H.M.; Writing—Review & Editing; M.d.M.C., E.A.-S. and A.S.-C.; Supervision, E.A.-S. and A.S.-C.

Funding: This work was supported by the International Cooperation Cell ICC06 under the Erasmus Mundus—Al Idrisi II programme "scholarship scheme for exchange and cooperation between Europe and North Africa". M.d.M.C. thanks the postdoctoral grant funded by the "Acción 6 del Plan de Apoyo a la Investigación de la Universidad de Jaén, 2017–2019".

Conflicts of Interest: The authors declare no conflict of interest.

References

1. Lamiales—Encyclopædia Britannica. Available online: https://www.britannica.com/plant/Lamiales (accessed on 20 June 2019).
2. Lim, T. Sesamum indicum. In *Edible Medicinal and Non-Medicinal Plants*; Springer: Dordrecht, The Netherlands, 2012; Volume 4, pp. 187–219.
3. Aboelsoud, N.H. Herbal medicine in ancient Egypt. *J. Med. Plants Res.* **2010**, *4*, 082–086.
4. Hassan, M.A. Studies on Egyptian sesame seeds (*Sesamum indicum* L.) and its products 1-Physicochemical analysis and phenolic acids of roasted Egyptian sesame seeds (*Sesamum indicum* L.). *World J. Dairy Food Sci.* **2012**, *7*, 195–201.
5. Grougnet, R.; Magiatis, P.; Laborie, H.; Lazarou, D.; Papadopoulos, A.; Skaltsounis, A.-L. Sesamolinol glucoside, disaminyl ether, and other lignans from sesame seeds. *J. Agric. Food Chem.* **2011**, *60*, 108–111. [CrossRef] [PubMed]
6. Grougnet, R.; Magiatis, P.; Mitaku, S.; Terzis, A.; Tillequin, F.; Skaltsounis, A.-L. New lignans from the perisperm of *Sesamum indicum*. *J. Agric. Food Chem.* **2006**, *54*, 7570–7574. [CrossRef] [PubMed]
7. Botelho, J.R.S.; Medeiros, N.G.; Rodrigues, A.M.C.; Araújo, M.E.; Machado, N.T.; Guimarães Santos, A.; Santos, I.R.; Gomes-Leal, W.; Carvalho Junior, R.N. Black sesame (*Sesamum indicum* L.) seeds extracts by CO_2 supercritical fluid extraction: Isotherms of global yield, kinetics data, total fatty acids, phytosterols and neuroprotective effects. *J. Supercrit. Fluid.* **2014**, *93*, 49–55. [CrossRef]
8. Dachtler, M.; van de Put, F.H.M.; v. Stijn, F.; Beindorff, C.M.; Fritsche, J. On-line LC-NMR-MS characterization of sesame oil extracts and assessment of their antioxidant activity. *Eur. J. Lipid Sci. Technol.* **2003**, *105*, 488–496. [CrossRef]
9. Namiki, M. Nutraceutical functions of sesame: A review. *Crit. Rev. Food Sci. Nutr.* **2007**, *47*, 651–673. [CrossRef] [PubMed]
10. Abouzed, T.K.; Contreras, M.d.M.; Sadek, K.M.; Shukry, M.; Abdelhady, D.H.; Gouda, W.M.; Abdo, W.; Nasr, N.E.; Mekky, R.H.; Segura-Carretero, A.; et al. Red onion scales ameliorated streptozotocin-induced diabetes and diabetic nephropathy in Wistar rats in relation to their metabolite fingerprint. *Diabetes Res. Clin. Pract.* **2018**, *140*, 253–264. [CrossRef] [PubMed]
11. FAO-Statitics. *Productions, Crops*; FAO-Statitics: Rome, Italy, 2014.

12. Mekky, R.H.; Contreras, M.d.M.; El-Gindi, M.R.; Abdel-Monem, A.R.; Abdel-Sattar, E.; Segura-Carretero, A. Profiling of phenolic and other compounds from Egyptian cultivars of chickpea (*Cicer arietinum* L.) and antioxidant activity: A comparative study. *RSC Adv.* **2015**, *5*, 17751–17767. [CrossRef]

13. Saura-Calixto, F.; Serrano, J.; Goni, I. Intake and bioaccessibility of total polyphenols in a whole diet. *Food Chem.* **2007**, *101*, 492–501. [CrossRef]

14. Ammar, S.; Contreras, M.d.M.; Belguith-Hadrich, O.; Segura-Carretero, A.; Bouaziz, M. Assessment of the distribution of phenolic compounds and contribution to the antioxidant activity in Tunisian fig leaves, fruits, skins and pulps using mass spectrometry-based analysis. *Food Funct.* **2015**, *6*, 3663–3677. [CrossRef] [PubMed]

15. The Dictionary of Natural Products Database. Available online: https://www.crcpress.com/go/the_dictionary_of_natural_products (accessed on 20 June 2019).

16. Singleton, V.; Rossi, J.A. Colorimetry of total phenolics with phosphomolybdic-phosphotungstic acid reagents. *Am. J. Enol. Viticult.* **1965**, *16*, 144–158.

17. Ammar, S.; Contreras, M.d.M.; Belguith-Hadrich, O.; Bouaziz, M.; Segura-Carretero, A. New insights into the qualitative phenolic profile of *Ficus carica* L. fruits and leaves from Tunisia using ultra-high-performance liquid chromatography coupled to quadrupole-time-of-flight mass spectrometry and their antioxidant activity. *RSC Adv.* **2015**, *5*, 20035–20050. [CrossRef]

18. Mohdaly, A.A.A.; Hassanien, M.F.R.; Mahmoud, A.; Sarhan, M.A.; Smetanska, I. Phenolics extracted from potato, sugar beet, and sesame processing by-products. *Int. J. Food Prop.* **2013**, *16*, 1148–1168. [CrossRef]

19. Ammar, S.; Contreras, M.d.M.; Gargouri, B.; Segura-Carretero, A.; Bouaziz, M. RP-HPLC-DAD-ESI-QTOF-MS based metabolic profiling of the potential *Olea europaea* by-product "wood" and its comparison with leaf counterpart. *Phytochem. Anal.* **2017**, *28*, 217–229. [CrossRef] [PubMed]

20. Suzuki, N.; Miyase, T.; Ueno, A. Phenylethanoid glycosides of *Sesamum indicum*. *Phytochemistry* **1993**, *34*, 729–732. [CrossRef]

21. Khaleel, A.E.S.; Gonaid, M.H.; El-Bagry, R.I.; Sleem, A.A.; Shabana, M. Chemical and biological study of the residual aerial parts of *Sesamum indicum* L. *J. Food Drug Anal.* **2007**, *15*, 249–257.

22. Eklund, P.C.; Backman, M.J.; Kronberg, L.Å.; Smeds, A.I.; Sjöholm, R.E. Identification of lignans by liquid chromatography-electrospray ionization ion-trap mass spectrometry. *J. Mass Spectrom.* **2008**, *43*, 97–107. [CrossRef]

23. Lee, J.; Choe, E. Extraction of lignan compounds from roasted sesame oil and their effects on the autoxidation of methyl linoleate. *J. Food Sci.* **2006**, *71*, C430–C436. [CrossRef]

24. Abu-Reidah, I.M.; Arráez-Román, D.; Segura-Carretero, A.; Fernández-Gutiérrez, A. Extensive characterisation of bioactive phenolic constituents from globe artichoke (*Cynara scolymus* L.) by HPLC–DAD-ESI-QTOF-MS. *Food Chem.* **2013**, *141*, 2269–2277. [CrossRef]

25. Fukuda, Y.; Osawa, T.; Namiki, M.; Ozaki, T. Studies on antioxidative substances in sesame seed. *Agric. Biol. Chem.* **1985**, *49*, 301–306.

26. Mekky, R.H. A Comparative phytochemical and biological studies on certain Egyptian varieties of *Cicer arietinum* Linn., Family Fabaceae. Ph. D. Thesis, Cairo University, Giza, Egypt, 2016.

27. Abu-Reidah, I.M.; Arráez-Román, D.; Quirantes-Piné, R.; Fernández-Arroyo, S.; Segura-Carretero, A.; Fernández-Gutiérrez, A. HPLC–ESI-Q-TOF-MS for a comprehensive characterization of bioactive phenolic compounds in cucumber whole fruit extract. *Food Res. Int.* **2012**, *46*, 108–117. [CrossRef]

28. Gómez-Romero, M.; Segura-Carretero, A.; Fernández-Gutiérrez, A. Metabolite profiling and quantification of phenolic compounds in methanol extracts of tomato fruit. *Phytochemistry* **2010**, *71*, 1848–1864. [CrossRef] [PubMed]

29. Kozukue, E.; Kozukue, N.; Tsuchida, H. Identification and changes of phenolic compounds in bamboo shoots during storage at 20 °C. *J. Jpn. Soc. Hortic. Sci.* **1998**, *67*, 805–811. [CrossRef]

30. Contreras, M.d.M.; Algieri, F.; Rodriguez-Nogales, A.; Gálvez, J.; Segura-Carretero, A. Phytochemical profiling of anti-inflammatory *Lavandula* extracts via RP-HPLC-DAD-QTOF-MS and -MS/MS: Assessment of their qualitative and quantitative differences. *Electrophoresis* **2018**, *39*, 1284–1293. [CrossRef]

31. Herzog, K.; IJlst, L.; van Cruchten, A.G.; van Roermund, C.W.T.; Kulik, W.; Wanders, R.J.A.; Waterham, H.R. An UPLC-MS/MS assay to measure glutathione as marker for oxidative stress in cultured cells. *Metabolites* **2019**, *9*, 45. [CrossRef] [PubMed]

32. Sahin, S.; Elhussein, E.A.A. Assessment of sesame (*Sesamum indicum* L.) cake as a source of high-added value substances: From waste to health. *Phytochem. Rev.* **2018**, *17*, 691–700. [CrossRef]

33. Janu, C.; Kumar, D.R.S.; Reshma, M.V.; Jayamurthy, P.; Sundaresan, A.; Nisha, P. Comparative study on the total phenolic content and radical scavenging activity of common edible vegetable oils. *J. Food Biochem.* **2014**, *38*, 38–49. [CrossRef]

34. Ross, I.A. *Medicinal Plants of the World*; Humana Press: New Jersey, USA, 2005; Volume 3, pp. 1–623.

35. Elgindi, M.; Ayoub, N.; Milad, R.; Mekky, R. Antioxidant and cytotoxic activities of Cuphea hyssopifolia Kunth (Lythraceae) cultivated in Egypt. *J. Pharmacogn. Phytochem.* **2012**, *1*, 67–77.

36. Mekky, R.H.; Fayed, M.R.; El-Gindi, M.R.; Abdel-Monem, A.R.; Contreras, M.d.M.; Segura-Carretero, A.; Abdel-Sattar, E. Hepatoprotective effect and chemical assessment of a selected Egyptian chickpea cultivar. *Front. Pharmacol.* **2016**, *7*, 1–9. [CrossRef]

37. Loizzo, M.R.; Sicari, V.; Pellicanò, T.; Xiao, J.; Poiana, M.; Tundis, R. Comparative analysis of chemical composition, antioxidant and anti-proliferative activities of Italian *Vitis vinifera* by-products for a sustainable agro-industry. *Food Chem. Toxicol.* **2019**, *127*, 127–134. [CrossRef] [PubMed]

38. Sok, D.E.; Cui, H.S.; Kim, M.R. Isolation and boactivities of furfuran type lignan compounds from edible plants. *Recent Pat. Food Nutr. Agric.* **2009**, *1*, 87–95. [CrossRef] [PubMed]

39. Suja, K.P.; Jayalekshmy, A.; Arumughan, C. Antioxidant activity of sesame cake extract. *Food Chem.* **2005**, *91*, 213–219. [CrossRef]

40. Materska, M. Flavone C-glycosides from *Capsicum annuum* L.: Relationships between antioxidant activity and lipophilicity. *Eur. Food Res. Technol.* **2015**, *240*, 549–557. [CrossRef]

41. Zhou, L.; Lin, X.; Abbasi, A.M.; Zheng, B. Phytochemical contents and antioxidant and antiproliferative activities of selected black and white sesame seeds. *Biomed. Res. Int.* **2016**, *2016*, 8495630. [CrossRef] [PubMed]

Article

Integrated Process for Sequential Extraction of Bioactive Phenolic Compounds and Proteins from Mill and Field Olive Leaves and Effects on the Lignocellulosic Profile

María del Mar Contreras [1,*], Antonio Lama-Muñoz [1], José Manuel Gutiérrez-Pérez [1,2], Francisco Espínola [1,2], Manuel Moya [1,2], Inmaculada Romero [1,2] and Eulogio Castro [1,2]

[1] Department of Chemical, Environmental and Materials Engineering, University of Jaén, Campus Las Lagunillas, 23071 Jaén, Spain; alama@ujaen.es (A.L.-M.); jmgp0014@red.ujaen.es (J.M.G.-P.); fespino@ujaen.es (F.E.); mmoya@ujaen.es (M.M.); iromero@ujaen.es (I.R.); ecastro@ujaen.es (E.C.)

[2] Center for Advanced Studies in Energy and Environment, University of Jaén, Campus Las Lagunillas, 23071 Jaén, Spain

[*] Correspondence: mcgamez@ujaen.es, mmcontreras@ugr.es or mar.contreras.gamez@gmail.com; Tel.: +34-953-21-27-99

Received: 25 September 2019; Accepted: 27 October 2019; Published: 29 October 2019

Abstract: The extraction of bioactive compounds in a biorefinery context could be a way to valorize agri-food byproducts, but there is a remaining part that also requires attention. Therefore, in this work the integrated extraction of phenolic compounds, including the bioactive oleuropein, and proteins from olive mill leaves was addressed following three schemes, including the use of ultrasound. This affected the total phenolic content (4475.5–6166.9 mg gallic acid equivalents/100 g), oleuropein content (675.3–1790.0 mg/100 g), and antioxidant activity (18,234.3–25,459.0 μmol trolox equivalents/100 g). No effect was observed on either the protein recovery or the content of sugars and lignin in the extraction residues. Concerning the recovery of proteins, three operational parameters were evaluated by response surface methodology. The optimum (63.1%) was achieved using NaOH 0.7 M at 100 °C for 240 min. Then, the selected scheme was applied to olive leaves from the field, observing differences in the content of some of the studied components. It also changed the lignocellulosic profile of the extraction residues of both leaf types, which were enriched in cellulose. Overall, these results could be useful to diversify the valorization chain in the olive sector.

Keywords: bioactive compounds; biorefinery; oleuropein; olive leaves; phenolic compounds; vegetable protein; ultrasound

---

1. Introduction

Olive tree cultivation is growing worldwide; the total area harvested was 10.8 million ha in 2017, which is three million more than in 1997 [1]. Thus, in addition to the main product, olive oil, high amounts of byproducts are generated. In particular, olive leaves (≈20%–25% by weight) are firstly generated during the tree pruning process and, secondly, in the mill leaves and thin branches (olive mill leaves) (≈4%–10% by weight) are separated together from olives using a blower machine [2–5]. This means that, for example, a hectare of olives trees could generate around 300–750 kg of olive leaves and 250 kg of olive mill leaves [3–6], or even more. These proportions may vary depending on the tree age, growing conditions, crop production, pruning intensity, local pruning practices, etc. [5,6]. Despite these large quantities, their industrial applications are still limited. In the worst scenario, these byproducts are burnt [6] and thereby contributing towards the emission of greenhouse gases.

Alternatively, olive leafy byproducts can be potential natural resources for obtaining valuable phytochemicals. Among them, oleuropein has revealed pharmacological potential in itself and as a starting material to develop new bioactive compounds [5,7,8]. Moreover, oleuropein contains a hydroxytyrosol moiety. Hydroxytyrosol and its derivatives (e.g., oleuropein complex and tyrosol) are the basis of the health claim on olive oil polyphenols approved by the Commission Regulation (EU) No. 432/2012, i.e., "olive oil polyphenols contribute to the protection of blood lipids from oxidative stress" [9]. In addition to the interest that oleuropein may cause, olive leaves extracts can provide a basis for the formulation of functional ingredients since a wide spectrum of bioactivities has been reported [5]. Similarly, olive mill leaves have antioxidant and antibacterial properties [2], but little is known about the oleuropein content.

Moreover, as food additives, olive leaves' extracts may counteract the loss of oil quality and enhance the stability of edible oils [10].

In this context, obtaining extracts rich in functional plant phytochemicals can be addressed to valorize agroindustrial byproducts [11], but generally the yield of extraction of phenolic compounds is low. This means that there is a large remaining fraction that can be applied for other purposes. Instead of a stand-alone process, the process based on the principles of biorefinery would increase the profitability [12]. This means that olive leafy byproducts can be complementarily used as feedstock for the production of second generation bioethanol from their sugar fraction [4,12]. Another unexplored fraction is proteins. Vegetable proteins can be extracted and used as such or in the form of hydrolyzates, with adequate nutritional and techno-functional properties [13–15]. New sources of usable protein could help to alleviate the global feed protein crisis [16]. Alkaline extraction is commonly used to recover plant proteins [13,15], but more studies are required to give new insights into the operational requirements when applied to leafy byproducts.

Therefore, the objective of this study was to integrate the sequential extraction of phenolic compounds, including the valuable compound oleuropein, and proteins from olive mill leaves and olive leaves from field. For that, maceration and ultrasound-assisted extraction of phenolic compounds was performed, while alkaline extraction was optimized via response surface methodology (RSM) to recover proteins and establish crucial factors affecting this step. Moreover, the residues obtained after extraction were characterized in terms of sugars and lignin since it can be valuable for other uses under a biorefinery approach.

2. Materials and Methods

2.1. Chemicals and Standards

For extraction, ethanol and sodium hydroxide (NaOH) were obtained from Sigma-Aldrich (St. Louis, MO, USA) and VWR Chemicals (Radnor, PA, USA), respectively. Acetone, methanol and acetonitrile were purchased from PanReac AppliChem (Barcelona, Spain). Folin and Ciocalteu's phenol reagent, sodium carbonate, 2,2'-azinobis(3-ethylbenzothiazoline-6-sulfonate) (ABTS), 6-hydroxy-2,5,7,8-tetramethylchroman-2-carboxylic acid (trolox), and standards of oleuropein, luteolin 7-*O*-glucoside and gallic acid were purchased from Sigma-Aldrich. Ultrapure water was obtained by a Milli-Q system (Millipore, Bedford, MA, USA).

2.2. Samples

Olive mill leaves from 'Picual' olive trees were collected in 2016 from the olive mill "SCA Unión Oleícola Cambil" (Jaén, Spain). Moreover, olive leaves were picked randomly in 2018 from olive tree leaves ('Picual') located in the Campus "Las Lagunillas" (University of Jaén). Leaves were washed with tap water, air-dried, and stored in a dry place until use. Just before starting the extraction process, both samples were ground (particle size around 1 mm) with an Ultra Centrifugal Mill ZM 200, Retsch (Haan, Germany).

2.3. Chemical Composition of Leafy Byproducts and Extraction Residues

The moisture and ash contents were determined according to the standard National Renewable Energy Laboratory (NREL) procedure [17]. According to the aforementioned procedure and after acid hydrolysis, carbohydrates were determined by high-performance liquid chromatography (HPLC) and lignin by gravimetric analysis. Acid soluble lignin was determined at 205 nm and a coefficient of extinction of 110 L/g cm was used [18]. The cellulose content was estimated from the glucose using an anhydro correction of 0.90 and hemicellulose from the other sugars using an anhydro correction of 0.90 and 0.88 for hexoses and pentoses, respectively [4,19].

2.4. Ethanolic Extraction

Extraction of phenolic compounds was based on the procedure of Ammar et al. [20], with some modifications. Briefly, olive leafy samples were extracted at 1:20 of solid-to-liquid ratio of initial weight using ethanol. Each sample was placed in a test tube, sonicated (40 kHz) (Ultrasons, J.P. Selecta, Barcelona, Spain) for 30 min at room temperature and centrifuged at 4000 rpm (Herolab, Wiesloch, Germany) for 15 min. Then, the supernatants were collected. For analysis, samples were filtered with a syringe filter (nylon, 0.45 μm pore size) (SinerLab Group, Madrid, Spain) and stored at 20 °C until analysis. Moreover, a portion of the extracts (15 mL) were oven-dried at 40 °C until constant weight. A control without sonication was also performed.

2.5. Alkaline Extraction

Alkaline extraction was initially performed at pH 9 in a bath (JULABO GmbH, Seelbach, Germany) at 60 °C and under agitation during 125 min. For that, NaOH at 0.03 M was added to olive mill leaves at a solid-to-liquid ratio of 1:10. These conditions were selected to be in the range of those reported in literature [13].

Then, the extraction conditions were optimized by RSM and the protein recovery was evaluated. The effect of NaOH concentration, extraction time and temperature were tested at three experimental levels using a central composite design (CCD) (2^3 + star, face centered). A total of 18 assays were carried out in randomized run order: eight points of a full factorial design (combination of levels 1 and −1), six star points, and four center points to estimate the experimental error. The assays were firstly performed at: i) mild conditions: pH 6–9 (i.e., NaOH concentration from 0.008 to 0.1 M); time, 10–240 min; temperature, 40–80 °C, and then ii) using strong conditions: NaOH concentration, 0.1–0.7 M; time, 10–240 min; temperature, 60–100 °C. The goodness of fit of the model was evaluated by the coefficient of determination (R^2), the lack of fit, and the residual standard deviation (RSD). The extraction at the optimum conditions were applied to olive leaves from mill and field and three repetitions were performed for each type of leaves.

In all cases, after subsequent centrifugation, which was performed at 4000 rpm (Herolab) for 15 min, supernatants were collected and stored at −20 °C until further analysis. Moreover, a portion of the extracts (15 mL) were oven-dried at 40 °C until constant weight.

2.6. Total Phenol Content (TPC) Method

The TPC was determined by a colorimetric assay using Folin–Ciocalteu reagent in 96-well polystyrene microplates, according to Mekky et al. [21]. A Bio-Rad iMarkTM microplate absorbance reader (Hercules, CA, USA) was employed. The absorbance was measured after incubation for 2 h in dark and compared with a calibration curve of gallic acid (25 to 300 μg/mL, $R^2 > 0.99$). The results were expressed as gallic acid equivalents (GAE).

2.7. Trolox Equivalent Antioxidant Capacity (TEAC) Assay

The TEAC assay was performed using the aforementioned microplate reader and following the procedure described by [21]. ABTS radical was produced by reacting ABTS with 2.45 mM potassium persulfate. The mixture was kept in dark at room temperature for 24 h and then diluted with water till reaching an absorbance value of 0.70 (± 0.02) at 734 nm. Afterwards, this solution and the extract (appropriately diluted) were mixed in the proportion 10:1 (*v/v*) and the absorbance measured. Absorbance readings were compared to a standard calibration curve of trolox (6 to 330 µM, $R^2 > 0.99$) and the results expressed as trolox equivalents (TE). Moreover, caffeic acid was used as a positive control (TEAC value ≈ 1.4 ± 0.1 mmol TE/mmol of compound).

2.8. Reversed-Phase (RP)-HPLC Analyses

The ethanolic extracts were analyzed using RP-HPLC coupled to UV. For that, a Shimadzu Prominence UFLC device was used, which was equipped with a DGU-20A5 degasser, LC-20AD quaternary pump, SIL-20AC HT auto sampler, SPD-M20A diode array detector and CTO-10AS VP column oven. A BDS HYPERSIL C18 column (290 mm × 4.6 mm, 5 µm particle size, Thermo Fisher Scientific Inc., Waltham, USA) was applied to separate the phenolic compounds. The mobile phase consisted of ultrapure water/0.2% orthophosphoric acid (solvent A), methanol (solvent B), and acetonitrile (solvent C) with an initial composition of 96/2/2 (*v/v/v*). A gradient elution at a flow rate of 1.0 mL/min and 30 °C was performed according to [22]. The obtained extracts were directly injected (20 µL) and the detection was performed in the UV range from 190 to 350 nm. Finally, for quantification, calibration curves at 280 nm were prepared with standards (from 2.5 to 1000 mg/L). The curves ($R^2 > 0.99$) were $y = 30{,}405x - 113{,}090$ for luteolin 7-*O*-glucoside and $y = 5591x + 11{,}911$ for oleuropein.

Additionally, RP-HPLC-MS and –MS^2 (working in automatic mode) was used to confirm the identity of the compounds. This analysis was performed on an Agilent 1100 HPLC System (Agilent Technologies, Waldbron, Germany) connected on-line to an Esquire 6000 ion trap (Bruker, Bremen, Germany). A linear gradient of solvent B (acetonitrile with formic acid, 0.1%, *v/v*) in A (water with formic acid, 0.1%, *v/v*) at a flow rate of 0.5 mL/min was applied according to Ammar et al. [20]. The column was a C18 Kinetex (2.1 × 50 mm, 2.7 µm) (Phenomenex, Barcelona, Spain) and the injection volume was 10 µL. Spectra were recorded over the mass-to-charge (*m/z*) range of 100–1200 in the negative ionization mode. Auto MS/MS analyses were performed at 0.6 *V*. About 4 spectra were averaged in the MS analyses and about 2 spectra in the MS/MS analyses. The data were processed using DataAnalysis (version 4.0) from Bruker.

2.9. Protein Content

The crude protein content of the byproducts was determined from the nitrogen content obtained by elemental analysis (TruSpec Micro, Leco, St. Joseph, MI, USA), applying a conversion factor of 6.25. The determination of the soluble protein was based on the Bradford assay, using a commercial kit from Bio-Rad. The absorbance was measured at 595 nm using the aforementioned colorimeter and bovine serum albumin (BSA) was used as standard for quantification to build a calibration curve up to 740 µg/mL ($R^2 > 0.99$). The protein recovery (%) was estimated as the ratio of protein content in the supernatant to the protein content of the byproducts.

2.10. Sodium Dodecyl Sulphate (SDS)-Polyacrylamide Gel Electrophoresis (PAGE)

For SDS–PAGE analysis, 100 µL of protein extract were precipitated by adding 400 µL of acetone for 20 min at cold conditions. The proteins were collected by centrifugation at 10,000× *g* for 10 min at 4 °C, and the resulting pellet was dissolved in 50 µL of Laemmli sample buffer containing 5% (*v/v*) 2-mercaptoethanol, according to [23]. In order to determine the molecular weight of the extracted protein products, their separation was carried out on Mini-PROTEAN® TGX™ Precast Gels (Bio-Rad).

Electrophoresis was performed at constant voltage (200 *V*) using Tris/Glycine/SDS buffer (Bio-Rad) as running buffer. Then, gels were stained during 1.5 h with Coomassie Brilliant Blue R-250 staining solution (Bio-Rad). Finally, gels were washed with a solution composed of water/methanol/acetic acid (60%:40%:10%, *v/v*) overnight. The molecular mass markers Precision Plus Protein™ Standard Unstained (10–250 kDa) (Bio-Rad) were used.

2.11. Sugar and Sugar Alcohol Analysis

Alkaline extracts were acidified using HCl 2 M (till pH around 3.5) and centrifuged as in Section 2.5. A portion of the supernatants was filtered (nylon, 0.45 μm pore size; SinerLab Group) and analyzed using an ICSep ICE-COREGEL-87H3 column (Transgenomic, Inc., Omaha, NE, USA) according to Martínez-Patiño et al. [24] and other portion was subjected to acid hydrolysis at 120 °C and analyzed as in 2.3.

2.12. Statistical Analysis

Statgraphics Centurion (StatPoint Technologies, Inc., Warrenton, VA, USA) was used to build the response surface experimental design and to obtain Pareto charts, which were used to summarize graphically and display the relative importance of each factor. One-way analysis of variance (ANOVA) followed by the least significant difference (LSD) multiple range test at the 0.05 significance level were also performed using the aforementioned software. The data are expressed as mean $\pm$ SD ($n = 3$).

3. Results and Discussion

3.1. Raw Composition of Leafy Byproducts

Table 1 shows the chemical composition of the olive byproducts after conditioning (drying and milling). Differences were found between both leaves types in terms of protein, cellulose (estimated as glucose), hemicellulose, lignin, ash, and mannitol ($p < 0.05$). In this regard, the hemicellulosic sugars of olive leafy biomass are mainly composed of xylose, arabinose, and galactose, which could come from xylans, arabinans, and galactans, respectively [25]. Although, other authors suggest that arabinans and galactans appear to be part of pectins, at least in the initial synthesis [26].

Concerning the nitrogen content in lignin, it could be derived from complexes formed between proteinaceous materials and lignin [27], was similar for both byproducts ($p = 0.09$), but there were differences between the percentage of acid insoluble protein with respect to the total protein content ($p < 0.05$), i.e., above 29% and 16% in olive mill leaves and olive leaves, respectively. Among other factors, all these differences could be explained by its primary origin since olive mill leaves consist mainly of olive leaves but mixed with a small amount of fine wood from small tree branches (<0.5 cm). As commented before, this byproduct is generated during olive harvesting and separated from olives using pneumatic separation systems in the mill.

Table 1. Chemical composition of olive mill leaves (OML) and olive leaves (OL) [1].

Component (%)	OML	OL
Crude protein	8.10 ± 0.38 [1]	9.34 ± 0.35 [1]
Cellulose (as glucose)	11.73 ± 0.14 [1]	15.84 ± 0.29 [1]
Hemicellulose	7.92 ± 0.05 [1,2]	8.62 ± 0.10 [1,2]
Mannitol	0.76 ± 0.02 [1]	2.81 ± 0.03 [1]
Acid soluble lignin	5.40 ± 0.09 [1]	7.46 ± 0.37 [1]
Acid insoluble lignin	35.16 ± 0.05 [1,3]	28.85 ± 1.05 [1,3]
N in acid-insoluble lignin	1.06 ± 0.11 [1]	0.82 ± 0.04 [1]
Ash	10.04 ± 0.08 [1]	6.24 ± 0.06 [1]

[1] Dry basis. [2] Composed of arabinose, xylose, and galactose. [3] With N.

3.2. Evaluation of the Extractions Schemes on Phenolic Compounds, Proteins, Sugars, and Lignin

3.2.1. Evaluation of the Extractions Schemes on Olive Mill Leaves

Preliminarily, three schemes for the sequential extraction of phenolic compounds (maceration or ultrasound-assisted extraction) and proteins (alkaline extraction) from olive mill leaves were evaluated: (Scheme 1) maceration, as control, followed by alkaline extraction; (Scheme 2) alkaline extraction followed by ultrasound-assisted extraction; and (Scheme 3) ultrasound-assisted extraction followed by alkaline extraction. For phenolic extraction, ethanol was selected as solvent since it has several advantages: among others, it is reusable, nontoxic with food grade status [27], as well as a potential biorefinery coproduct. Concerning alkaline extraction, mild conditions (initial pH 9; temperature, 60 °C; time, 125 min) were applied, according to those previously reported [13]. Table 1 shows the values for TPC, the content of the olive bioactives, oleuropein and luteolin 7-*O*-glucoside, and the antioxidant activity determined by TEAC.

Using ultrasound-assisted extraction (Scheme 3) to recover phenolic compounds and as first step, higher values for TPC, oleuropein content, luteolin 7-*O*-glucoside content and TEAC were obtained as compared to solely maceration (Scheme 1) (Table 2); between 4% and 34% higher. The use of ultrasound generally favors the extraction of phenolic compounds from plant materials, but this enhancement depends on the conditions used (including the device) and the biomass type [21,28]. In another context, when ultrasound-assisted extraction was performed after protein extraction (Scheme 2), the phenolic profile changed mainly quantitatively (Figure 1), and the amounts of oleuropein and luteolin 7-*O*-glucoside were lower (Table 2). Oleuropein could suffer thermal degradation during alkaline extraction following this scheme, caused by oxidation, cleavage of covalent bonds or enhanced oxidation reactions as suggested by [29]. Moreover, the use of alkaline conditions could modify oleuropein to give low active degradation products, as suggested by Soler-Rivas et al. [30]. On the contrary, the values for TPC and TEAC were the highest using this scheme (Scheme 2). This could be also related to the change in the phenolic profile that led to obtain more luteolin in this extract (Figure 1). This fact can explain these results taking into account the results by Benavente-García et al. [7], who reported the antioxidant activity of some olive phenolic compounds, including luteolin, luteolin 7-*O*-glucoside, and oleuropein, using this antioxidant assay. Thus, the phenolic composition and the antioxidant activity of the extracts can be modulated by the sequential extractions scheme applied. Nonetheless, the TEAC method is primarily governed by steric considerations of the radical, and the presence of hydrogen atom transfer-acting antioxidants, which react slowly in this system, could be underestimated [31].

Table 2. Total phenolic yield (%), total phenol content (TPC) (mg gallic acid equivalents/100 g), oleuropein (Ole) content (mg/100 g), luteolin 7-*O*-glucoside (L7G) content (mg/100 g) and antioxidant activity (μmol trolox equivalents/100 g) of the ethanolic extracts as well as protein recovery (%) after alkaline extraction obtained by three different sequential extractions schemes.

Byproduct [#]	Scheme [‡]	Phenolic Yield	TPC/Biomass	TPC/Extract	Ole/Biomass	Ole/Extract	L7G/Biomass	L7G/Extract	TEAC [#]/ Biomass	TEAC [#]/ Extract	Protein Recovery
OML	1	10.1 ± 0.2 [b]	454 ± 36 [b]	4476 ± 284 [a]	142 ± 3 [ab]	1395 ± 6 [ab]	31 ± 4 [ab]	309 ± 46 [a]	2024 ± 2 [b]	18,234 ± 139 [b]	11.2 ± 0.8 [a]
OML	2	11.7 ± 0.2 [a]	719 ± 89 [a]	6167 ± 847 [a]	79 ± 0.3 [b]	675 ± 6 [b]	28 ± 0.3 [b]	237 ± 6 [b]	3087 ± 143 [a]	25,459 ± 2809 [a]	13.7 ± 1.4 [a]
OML	3	11.7 ± 0.6 [aA]	585 ± 24 [abB]	4998 ± 65 [aB]	191 ± 45 [aB]	1790 ± 434 [aB]	38 ± 3 [aB]	338 ± 21 [aB]	2193 ± 69 [bB]	19,050 ± 101 [bB]	12.5 ± 0.8 [a]
OL	3	10.8 ± 0.8 [A]	1405 ± 99 [A]	13,108 ± 1877 [A]	1365 ± 6 [A]	12,694 ± 694 [A]	97 ± 1 [A]	903 ± 8 [A]	6364 ± 38 [A]	59,651 ± 6429 [A]	ND

[#] OML, olive mill leaves; OL, olive leaves; TEAC, trolox equivalent antioxidant capacity. [‡] Sequential extractions. Scheme 1: maceration with ethanol before protein extraction; Scheme 2: ethanolic extraction assisted by ultrasound after protein extraction; Scheme 3: ethanolic extraction assisted by ultrasound before protein extraction. In each column, data followed by the same minor letter are not statistically different from each other concerning the three schemes applied on OML, while data followed by the same capital letter are not statistically different from each other for the Scheme 3 applied in OML and OL (least significant difference test, $p < 0.05$).

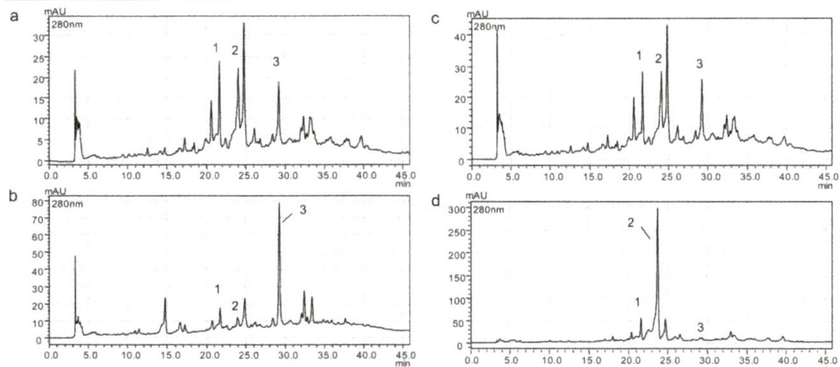

Figure 1. Chromatograms (280 nm) of ethanol extracts obtained by: (**a**) maceration of olive mill leaves (OML) before protein extraction (Scheme 1), (**b**) ultrasound-assisted extraction of OML after protein extraction (Scheme 2), (**c**) ultrasound-assisted extraction of OML before protein extraction (Scheme 3), and (**d**) ultrasound-assisted extraction of olive leaves before protein extraction (Scheme 3). (1) Luteolin 7-*O*-glucoside, (2) oleuropein, and (3) luteolin.

Concerning the recovery of proteins, there were no differences among the extraction schemes (Table 2). Although the study of Karki et al. [32] suggested that a pretreatment with ultrasound may enhance protein release from soy meal, the byproduct type, the ultrasonic device and the conditions applied were different. Moreover, our results suggested that the residues of extraction contained similar amount of solids, ash, acid-soluble lignin (ASL), acid-insoluble lignin (AIL), protein linked to AIL, cellulose, and hemicellulose (Figure 2). In all cases, the most susceptible fraction was lignin with only 60%–65% remained in the residue after extraction. This makes sense since alkaline pretreatments are used for removing lignin from the biomass in order to increase the accessibility and digestibility of cellulose for saccharification and transformation into biofuels [33].

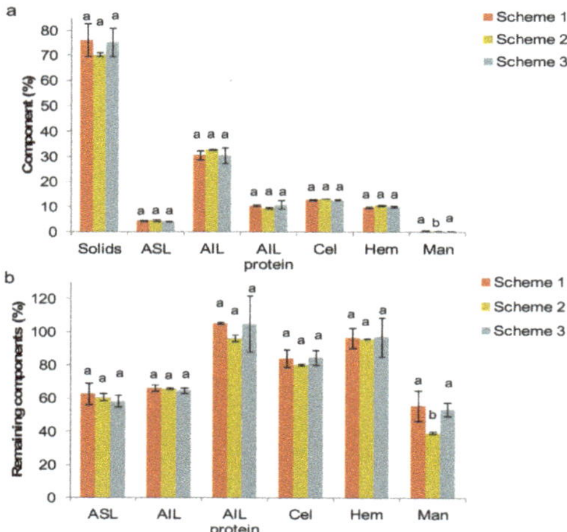

Figure 2. (**a**) Yield of solids (%) and content (%) of acid-soluble lignin (ASL), acid-insoluble lignin (AIL), protein in AIL, cellulose (Cel), hemicellulose (Hem), and mannitol (Man) in the remaining fraction from olive mill leaves (OML) after extraction using Schemes 1–3. (**b**) Corresponding recovery values (%) with respect to the initial amounts in OML.

3.2.2. Comparison between Olive Mill Leaves and Olive Leaves from Field

Overall, since oleuropein is a potential therapeutic molecule as commented before, the extraction of phenolic compounds was performed before alkaline extraction in subsequent experiments since the rest of constituents were not significantly affected by the extractions sequence followed. Nevertheless, if higher antioxidant activity is desired, protein extraction can be performed before phenolic extraction. If the use of ultrasound is not possible in the industrial scheme, the phenolic amounts in the extracts will be reduced only slightly.

Therefore, for comparison, olive leaves were also extracted by ultrasound-assisted extraction as first extraction step (Table 2). This extract presented higher TPC, oleuropein content, luteolin 7-O-glucoside content and antioxidant activity than that obtained from olive mill leaves. These differences could be explained by the fact that the composition of the byproducts is different. As commented before, olive mill leaves not only contain leaves but also present woody material. It seems that the content of oleuropein in olive wood is lower than in leaves [20,34,35], being absent in the wood of some cultivars [35]. Also, storage time and conditions could have affected the latter values, since olive leaves were picked fresh before extraction, while olive mill leaves were stored at room temperature as it was at the mill. In any case, olive mill leaves are a cheap and easily accessible source of oleuropein. This means that this leafy byproduct is localized and stored in the mill, being ready for utilization, but storage time and conditions should be further controlled in a future biorefinery based on the production of antioxidant extracts.

3.3. Evaluation of the Solubilization of Proteins by Mild Alkaline Conditions after Phenolic Extraction

Alkaline extraction is commonly employed to extract proteins from vegetable sources, but it has not been well explored in leafy byproducts [13,36]. Thus, RSM was applied to evaluate the effect of some parameters on the solubilization of proteins from olive mill leaves using mild alkaline-thermal conditions at a solid-to-liquid ratio of 1:10. Table S1 shows the experimental levels of the tested factors, i.e., initial pH from 6 to 9, extraction time from 10 to 240 min, and temperature from 40 to 80 °C, along with the results obtained for the response variable (protein recovery) and the yield.

Figure 3(a1) shows the Pareto chart of the standardized effect of each term on the protein recovery and its statistical significance at the 90% confidence level, while Figure 3(a2) shows the main effects plots. The Pareto chart (Figure 3(a1)) indicates that pH, temperature, and the interaction of both parameters had the strongest influence on the protein recovery, as well as this effect was positive. The rest of the variables, including the extraction time and the quadratic terms of the parameters, had no significant effects on the extraction recovery, hence they were eliminated from the model. In this way, the corresponding surface plot is depicted in Figure 3(a3). The model had an R^2 of 85.16%, the standard error of the estimate was 1.83 and the *p*-value of the lack-of-fit was 0.82 (Table 3).

The model proposed was:

$$Y = a_0 + \sum_{i=1}^{3} a_i\, X_i + \sum_{i=1}^{3} a_{ii}\, X_i^2 + \sum_{i \neq j = 1}^{3} a_{ij}\, X_i\, X_j \tag{1}$$

where Y is the response variable, a_0 is a constant, a_i, a_{ii}, and a_{ij} are the linear, quadratic, and interaction coefficients, respectively. The values of the coefficients are shown in Table 3. Using this model, the optimum conditions were: pH 12 (NaOH concentration of 0.1 M), 80 °C and 240 min. The predicted recovery value was 21.5%, which was similar to the experimental value (21.7% ± 2.3%). Moreover, these extraction conditions were applied to olive leaves, but the recovery of proteins was slightly lower at 15.5% ± 0.2%. In this regard, protein extractability depends on the byproduct type and composition [13,37]. Olive leaves contain more cellulose than olive mill leaves (Table 1) and cellulose may hamper the extractability of proteins [37], explaining at least in part our results.

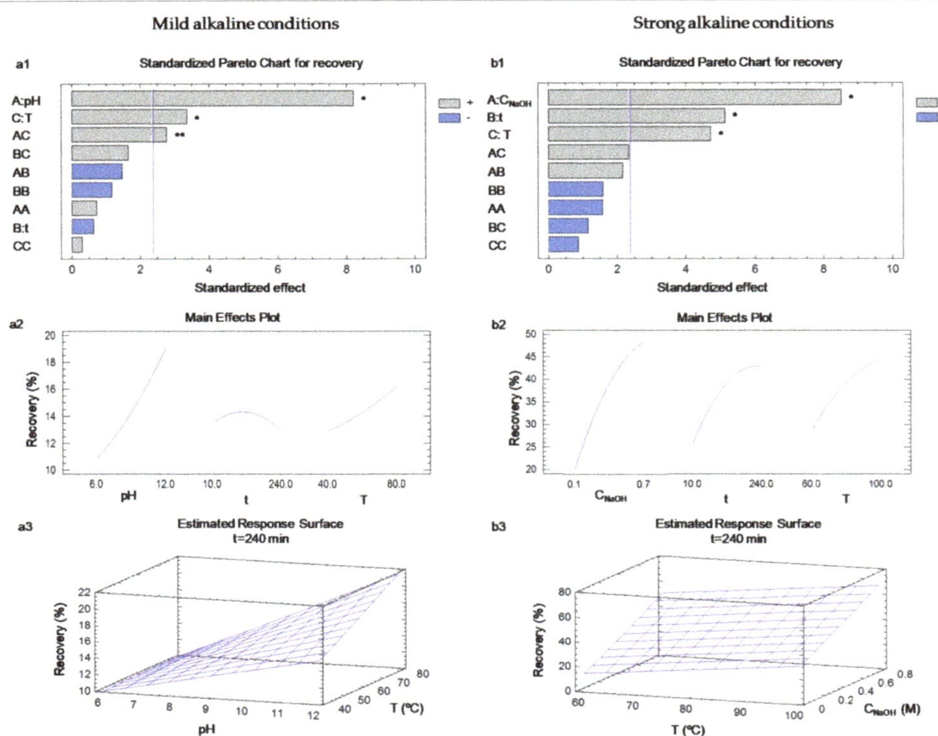

Figure 3. Pareto charts indicating the weight of each factor on the protein recovery and corresponding main effects plots using (**a1** and **a2**, respectively) mild and (**b1** and **b2**, respectively) strong alkaline conditions. The surface plots are represented take into account the significant factors: (**a3**) mild and (**b3**) strong alkaline conditions. C_{NaOH}, NaOH concentration; T, temperature; t, extraction time. * Significant at p-value < 0.05; ** significant at p-value < 0.1.

Table 3. Model equation coefficients and optimum conditions values for the recovery of proteins from olive mill leaves obtained using mild and strong alkaline extractions.

Equation Terms	Mild Alkaline Conditions	Strong Alkaline Conditions
Coefficients [1]		
a_0	10.785	−27.166
Linear		
a_1	−0.174 ***	47.100 ***
a_2	NS	0.074 **
a_3	−0.147 **	0.392 **
Interaction		
a_{12}	NS	NS
a_{13}	0.026 *	NS
a_{23}	NS	NS
Quadratic		
a_{11}	NS	NS
a_{22}	NS	NS
a_{33}	NS	NS

Table 3. *Cont.*

Equation Terms	Mild Alkaline Conditions	Strong Alkaline Conditions
R^2	0.85	0.69
Lack-of-fit	0.823	0.113
Optimum (estimated)	22.3	62.7
Optimum (experimental)	21.7 ± 2.3	63.1 ± 5.7
1 (pH/C_{NaOH})	12	0.7 M
2 (time)	240 min	240 min
3 (temperature)	80 °C	100 °C

[1] The factors were pH (1), time (2) and temperature (3) in the design for mild alkaline conditions and NaOH concentration (C_{NaOH}) (1), time (2) and temperature (3) in the design for strong alkaline conditions. NS, not significant; significant at *** $p < 0.01$; ** $0.01 < p < 0.05$; * $0.05 < p < 0.1$.

3.4. Evaluation of the Solubilization of Proteins by Strong Alkaline-thermal Conditions

Since temperature and pH (determined by the NaOH amount) were the most important factors in the former design, a new design was built to evaluate the effect of stronger NaOH concentration and temperature. Firstly, the effect of the amount of alkali added per solid (1–7 mmol NaOH/g of solid) on the solubilization of proteins was evaluated. For that, NaOH solutions from 0.1 to 0.7 M at a fixed solid-to-liquid ratio of 1:10 were tested, and as well a fixed value of NaOH 0.1 M at different solid-to-liquid ratio values (1:10–1:70). Figure 4 depicts that the use of higher amounts of alkali increased the amount of protein extracted, particularly when using concentrated NaOH solutions. This led to higher pH values (up to 13.3) than using the other way (up to 12.6). These results agree with those obtained by Zhang et al. [38], who reported that the amount of applied alkali is critical to extract proteins from leafy byproducts.

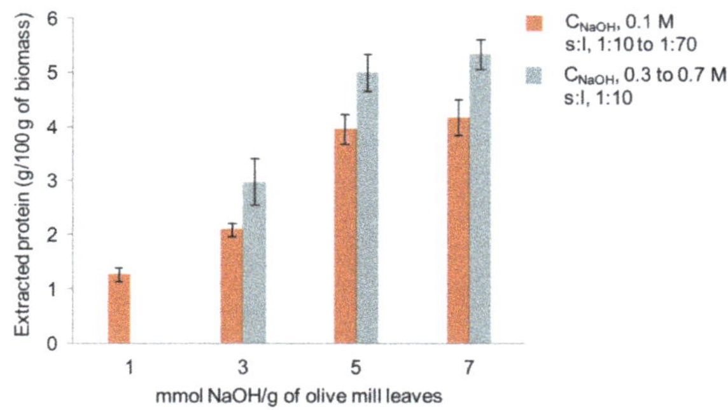

Figure 4. Amount of protein extracted using different ratio of alkali to solid, which was obtained using different NaOH concentration (C_{NaOH}) and solid-to-liquid ratio (s:l) values.

Secondly, taking into account the previous results, the solid-to-liquid ratio was fixed again to 1:10 to reduce the consumption of water and the NaOH concentration (0.1–0.7 M), the temperature (60–100 °C) and the extraction time (10–240 min) were optimized by using a CCD (Table 4). In this case, NaOH concentration, temperature and extraction time had the strongest influence on the protein recovery (p-value < 0.05), as well as this effect was positive (Figure 3(b1)). In this case, the interaction and quadratic terms had no significant effects on the protein recovery. The influence of these three operational parameters has also been reported in algae [39] and tea byproduct [38]. Furthermore, the use of high temperatures seems to be essential to extract proteins from leafy byproducts in agreement with Sari et al. [37].

Table 4. Protein recovery from olive mill leaves subjected to different alkaline-thermal treatments (strong conditions).

Assay No.	pH	NaOH Concentration (M)	Time (min)	Temperature (°C)	Protein Recovery (%)	Yield (Solids) (%)
1	13.2	0.4	125	80	36.1	40.0
2	13.6	0.7	240	60	43.4	57.4
3	13.5	0.7	240	100	55.8	66.5
4	12.2	0.1	240	100	15.5	24.4
5	12.3	0.1	240	60	14.4	23.4
6	12.2	0.1	10	100	13.6	23.2
7	13.3	0.4	10	80	13.6	32.1
8	13.4	0.4	125	80	38.4	42.2
9	13.1	0.7	10	60	17.1	50.3
10	13.7	0.7	125	80	51.3	60.1
11	12.5	0.1	10	60	10.0	39.2
12	12.7	0.1	125	80	16.7	21.2
13	12.7	0.4	125	100	53.5	50.0
14	13.2	0.7	10	100	43.9	63.3
15	13.3	0.4	125	80	36.8	42.6
16	13.4	0.4	240	80	54.3	49.3
17	12.8	0.4	125	60	19.0	41.6
18	13.2	0.4	125	80	50.9	46.1

Finally, the model was rebuilt considering only the significant variables and the surface plot is shown in Figure 3(b3). The new model explained almost 70% of the variability and the standard error of the estimate was 5.3%. Since the p-value for lack-of-fit in the ANOVA was greater than 0.05 (Table 3), the model appears to be adequate for the observed data at the 95.0% confidence level. Table 3 also details the coefficients for Equation (1) and the optimum conditions, which were obtained using NaOH 0.7 M at 100 °C for 240 min. The predicted recovery value was 62.7%, which is similar to the experimental value (63.1% ± 5.7%), i.e., ≈ 5 g/100 g of olive mill leaves. Finally, the optimum conditions were applied to olive leaves. The recovery value was 55.5% ± 4.3% (i.e., ≈ 5 g/100 g of OL); again it was slightly lower than that for olive mill leaves.

It should be noticed that a recovery value higher than 50% was obtained using the conditions assayed in the experiment 16 (NaOH 0.4 M, 80 °C for 240 min) (Table 4). Although this value is lower than that using the optimum conditions, the alkali and temperature requirements are lesser. This treatment was also applied in the subsequent experiments.

3.5. Characterization of the Protein Products by SDS-PAGE

The solubilized protein consisted of proteins partially hydrolyzed into peptides with molecular weight lower than 10 kDa (band B1), proteins/peptides closer to 10 kDa (band B2) and 100 kDa (band B3) (Figure S1). When using stronger thermal-alkaline conditions, bands B1 and B2 were more prominent, suggesting it may favor protein hydrolysis, in agreement with Fetzer et al. [14]. Nonetheless, Zhang et al. [40] reported that tea protein was not severely hydrolyzed after alkaline treatment at 95 °C, 0.1 M NaOH and a v/w of 40:1. Moreover, the band B4 (>250 kDa) was not well resolved and could be possibly formed by complexed proteins too large to enter the gel [15].

Some similar bands have been previously reported in olive leaves [22,41], while RuBisCO main subunit band at 55 kDa was not detected. This protein could be affected by hydrolysis reactions occurred under the conditions applied or complexation. Furthermore, all these protein bands were also observed in the alkaline extract from olive leaves, suggesting that the protein precursors are similar for both byproducts (Figure S2).

3.6. Characterization of the Residual Fraction after the Sequential Extraction of Phenolic Compounds and Protein

For a complete valorization of olive mill leaves, the remaining fraction obtained after the sequential extractions scheme was characterized (Figure 5). Under optimum alkaline extraction conditions (i.e., NaOH concentration, 0.7 M; temperature, 100 °C; time, 4 h) (Scheme 3″), the percentage of lignin was lower than using softer thermal-alkaline conditions, i.e., Scheme 3 (NaOH concentration, 0.03 M; temperature, 60 °C; time, 125 min) and scheme 3′ (NaOH concentration, 0.4 M; temperature, 80 °C; time, 4 h) (Figure 5a). This means that the chemical profiles are different from each other and with respect to the raw byproduct. Particularly, using Scheme 3″, the recovery of most components was lower, which could pass to the liquid phase as hydrolyzed forms (Figure 5b). Similarly, the latter extraction conditions also changed the chemical profile of the remaining fraction recovered from olive leaves compared to the raw byproduct. In this case, the chemical profile and the recovery values were similar to those of olive mill leaves, with the exception of AIL and mannitol.

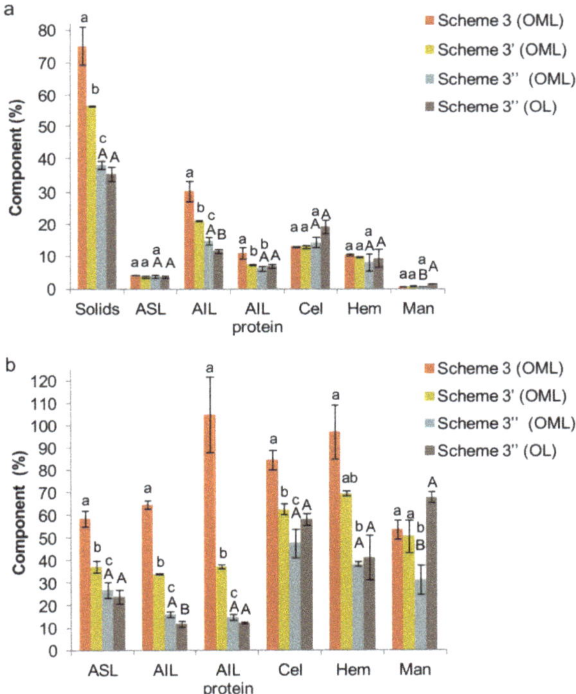

Figure 5. (**a**) Yield of solids (%) and content (%) of acid-soluble lignin (ASL), acid-insoluble lignin (AIL), protein in AIL, cellulose (Cel) and hemicellulose (Hem) and mannitol (Man) in the remaining fraction from olive mill leaves (OML) and olive leaves (OL) after phenolic extraction followed by alkaline extraction using NaOH 0.03 M, 60 °C, 125 min (Scheme 3), NaOH 0.4 M, 80 °C, 4 h (Scheme 3′), and NaOH 0.7 M, 100 °C, 4 h (Scheme 3″, optimum conditions). (**b**) Corresponding recovery values (%) with respect to the initial amounts in both byproducts.

In general, the lowest recovery value was found for lignin (both soluble and insoluble fractions) and AIL protein (or protein linked to lignin) (Scheme 3″), suggesting that these components were more solubilized than the others. It seems that an alkali treatment can attack mainly hydrolysable linkages in lignin, which cause a reduction in the degree of polymerization and disruption of the lignin structure of biomass [42]. This can lead to the breakage of lignin linkages, such as aryl–ether, ester and C–C bonds [43], as well as linkages between lignin–carbohydrate complexes [44], whose presence has been reported in different types of biomass [45]. Among the first type, β-*O*-4′ alkyl–aryl ethers are the most abundant lignin inter-unit linkages in olive tree pruning, which is composed of olive leaves, thin branches and wood [46]. Nonetheless, these authors suggested that its low syringyl/guaiacyl ratio and the presence of condensed structures (C–C) make it probably less reactive than other biomasses, and thus requiring higher amounts of alkali. Furthermore, the destabilization of the polymeric structure of the biomass could favor the release proteins linked to fibers, but literature information is scarce. In our case, the alkaline extraction also led to obtain cellulose and hemicellulose enriched fractions with higher ratios of sugars/lignin, which can be further valorized for obtaining biofuels as shown McIntosh and Vancov [47] for alkaline pretreated wheat straw.

Alternatively, the solubilization of lignin and its co-precipitation with proteins via acid precipitation (until ≈pH 3.5) can explain, at least in part, that the protein enrichment was modest as suggested other authors [13,48]. The protein content of acid precipitates was up to 24% (Scheme 3′). In this context, future studies should be addressed to separate proteins from solubilized lignin and sugars in the alkaline extracts since all these components are valuable, e.g., using enzymes and acids [48]. Concerning sugars and derivatives, mannitol (1.0–2.3 g/L), xylitol (0.4–1.1 g/L), and arabinose oligomers (21.5–35.8 g/L) were detected in the alkaline extracts. The formers have many applications as natural sweeteners and excipients in the food industry and pharmaceutical industries [4], while the latter could be useful as a prebiotics [49].

4. Conclusions

The following scheme could be applied to obtain phenolic compounds and proteins from olive leafy byproducts: ultrasound-assisted extraction of phenolic compounds to recover oleuropein followed by alkaline extraction of proteins. The amount of oleuropein extracted per 100 g of biomass was higher in olive leaves (≈1.4 g) than in olive mill leaves (≈0.2 g), while the extracted protein (≈5 g) amount was similar. If higher antioxidant activity is desired, protein extraction can be performed before phenolic extraction, also increasing the luteolin content. Furthermore, to increase the recovery of proteins from this leafy byproduct, strong alkaline-thermal conditions are required. Alkaline extraction led to changes in the residual lignocellulosic fraction, which was enriched in cellulose. Further studies are required to assess its viability for obtaining biofuels in biorefinery and to purify proteins and other interesting compounds, such as oligosaccharides.

Supplementary Materials: The following are available online at http://www.mdpi.com/2304-8158/8/11/531/s1, Table S1: Protein recovery from olive mill leaves subjected to different alkaline-thermal treatments (mild conditions). Figure S1: SDS-PAGE of protein extracts from olive mill leaves obtained at different extraction conditions, Figure S2: SDS-PAGE of protein extracts from olive mill leaves (OML) and olive leaves (OL) obtained using 0.7 M NaOH at 100 °C for 4 h.

Author Contributions: Conceptualization, M.d.M.C. and E.C.; methodology, M.d.M.C., J.M.G.-P., and A.L.-M.; software, M.d.M.C.; validation, M.d.M.C.; writing—original draft preparation, M.d.M.C.; writing—review and editing, M.d.M.C., F.E., M.M., I.R., and E.C.; supervision, M.d.M.C. and E.C.; funding acquisition, M.d.M.C., I.R., and E.C.

Funding: "Ministerio de Economía y Competitividad" (Spain), reference project ENE2017-85819-C2-1-R, including FEDER funds. Financial support from "Agencia Estatal de Investigación" and "Fondo Europeo de Desarrollo Regional". The authors are also grateful for the postdoctoral grants funded by the "Acción 6 del Plan de Apoyo a la Investigación de la Universidad de Jaén, 2017–2019".

Acknowledgments: The technical and human support provided by CICT of the University of Jaén (UJA, MINECO, Junta de Andalucía, FEDER) is gratefully acknowledged.

Conflicts of Interest: The authors declare no conflict of interest.

References

1. FAOSTAT. Available online: http://www.fao.org/faostat/en/#data (accessed on 13 February 2019).
2. Gullón, B.; Gullón, P.; Eibes, G.; Cara, C.; De Torres, A.; López-Linares, J.C.; Ruiz, E.; Castro, E. Valorisation of olive agro-industrial by-products as a source of bioactive compounds. *Sci. Total Environ.* **2018**, *645*, 533–542. [CrossRef] [PubMed]
3. Romero-García, J.M.; Niño, L.; Martínez-Patiño, C.; Álvarez, C.; Castro, E.; Negro, M.J. Biorefinery based on olive biomass. State of the art and future trends. *Bioresour. Technol.* **2014**, *159*, 421–432. [CrossRef] [PubMed]
4. Romero-García, J.M.; Lama-Muñoz, A.; Rodríguez-Gutiérrez, G.; Moya, M.; Ruiz, E.; Fernández-Bolaños, J.; Castro, E. Obtaining sugars and natural antioxidants from olive leaves by steam-explosion. *Food Chem.* **2016**, *210*, 457–465. [CrossRef]
5. Talhaoui, N.; Taamalli, A.; Gómez-Caravaca, A.M.; Fernández-Gutiérrez, A.; Segura-Carretero, A. Phenolic compounds in olive leaves: Analytical determination, biotic and abiotic influence, and health benefits. *Food Res. Int.* **2015**, *77*, 92–108. [CrossRef]
6. Ruiz, E.; Romero-García, J.M.; Romero, I.; Manzanares, P.; Negro, M.J.; Castro, E. Olive-derived biomass as a source of energy and chemicals. *Biofuel. Bioprod. Biorefin.* **2017**, *11*, 1077–1094. [CrossRef]
7. Benavente-García, O.; Castillo, J.; Lorente, J.; Ortuño, A.; Del Rio, J.A. Antioxidant activity of phenolics extracted from *Olea europea* L. leaves. *Food Chem.* **2000**, *68*, 457–462. [CrossRef]
8. Cavaca, L.A.S.; Alfonso, C.A.M. Oleuropein: A valuable bio-renewable synthetic building block. *Eur. J. Org. Chem.* **2018**, *2018*, 581–589. [CrossRef]
9. Commission Regulation (EU) N° 432/2012 of 16 May 2012 establishing a list of permitted health claims made on foods, other than those referring to the reduction of disease risk and to children's development and health. *Off. J. Eur. Union* **2012**, *136*, 1–40.
10. Ammar, S.; Kelebek, H.; Zribi, A.; Abichou, M.; Selli, S.; Bouaziz, M. LC-DAD/ESI-MS/MS characterization of phenolic constituents in Tunisian extra-virgin olive oils: Effect of olive leaves addition on chemical composition. *Food Res. Int.* **2017**, *100*, 477–485. [CrossRef]
11. Mekky, R.H.; Abdel-Sattar, E.; Segura-Carretero, A.; Contreras, M.d.M. Phenolic Compounds from sesame cake and antioxidant activity: A new insight for agri-food residues' significance for sustainable development. *Foods* **2019**, *8*, 432. [CrossRef]
12. Solarte-Toro, J.C.; Romero-García, J.M.; Susmozas, A.; Ruiz, E.; Castro, E.; Cardona-Alzate, C.A. Techno-economic feasibility of bioethanol production via biorefinery of olive tree prunings (OTP): optimization of the pretreatment stage. *Holzforschung* **2019**, *73*, 3–13. [CrossRef]
13. Contreras, M.d.M.; Lama-Muñoz, A.; Gutiérrez-Pérez, J.M.; Espínola, F.; Moya, M.; Castro, E. Protein extraction from agri-food residues for integration in biorefinery: Potential techniques and current status. *Bioresour. Technol.* **2019**, *280*, 459–477. [CrossRef] [PubMed]
14. Fetzer, A.; Herfellner, T.; Stäbler, A.; Menner, M.; Eisner, P. Influence of process conditions during aqueous protein extraction upon yield from pre-pressed and cold-pressed rapeseed press cake. *Ind. Crop. Prod.* **2018**, *112*, 236–246. [CrossRef]
15. Connolly, A.; Piggott, C.O.; Fitzgerald, R.J. Characterisation of protein-rich isolates and antioxidative phenolic extracts from pale and black brewers' spent grain. *Int. J. Food Sci. Technol.* **2013**, *48*, 1670–1681. [CrossRef]
16. Lei, X.G. Sustaining the future of animal feed protein. *Ind. Biotechnol.* **2018**, *14*, 74–76. [CrossRef]
17. Determination of Structural Carbohydrates and Lignin in Biomass. Laboratory Analytical Procedure (LAP). Available online: https://www.nrel.gov/docs/gen/fy13/42618.pdf (accessed on 13 February 2019).
18. Hatfield, R.; Fukushima, R.S. Can Lignin Be Accurately Measured? *Crop Sci.* **2005**, *45*, 832–839. [CrossRef]
19. Ammar, S.; Contreras, M.d.M.; Gargouri, B.; Segura-Carretero, A.; Bouaziz, M. RP-HPLC-DAD-ESI-QTOF-MS based metabolic profiling of the potential Olea europaea by-product "wood" and its comparison with leaf counterpart. *Phytochem. Anal.* **2017**, *28*, 217–229. [CrossRef]
20. Mekky, R.H.; Contreras, M.d.M.; Roshdi El-Gindi, M.; Abdel-Monem, A.R.; Abdel-Sattar, E.; Segura-Carretero, A. Profiling of phenolic and other compounds from Egyptian cultivars of chickpea (*Cicer arietinum* L.) and antioxidant activity: A comparative study. *RSC Adv.* **2015**, *5*, 17751–17767. [CrossRef]
21. Determination of Biophenols in Olive Oils by HPLC. Available online: http://www.internationaloliveoil.org/documents/viewfile/4141-met29eng (accessed on 13 February 2019).

22. Vergara-Barberán, M.; Lerma-García, M.J.; Herrero-Martínez, J.M.; Simó-Alfonso, E.F. Use of an enzyme-assisted method to improve protein extraction from olive leaves. *Food Chem.* **2015**, *169*, 28–33. [CrossRef]

23. Martínez-Patiño, J.C.; Romero, I.; Ruiz, E.; Cara, C.; Romero-García, J.M.; Castro, E. Design and optimization of sulfuric acid pretreatment of extracted olive tree biomass using response surface methodology. *BioResources* **2017**, *12*, 1779–1797. [CrossRef]

24. Zhao, X.; Liu, D. Chemical and thermal characteristics of lignins isolated from Siam weed stem by acetic acid and formic acid delignification. *Ind. Crop. Prod.* **2010**, *32*, 284–291. [CrossRef]

25. Cara, C.; Ruiz, E.; Carvalheiro, F.; Moura, P.; Ballesteros, I.; Castro, E.; Gírio, F. Production, purification and characterisation of oligosaccharides from olive tree pruning autohydrolysis. *Ind. Crop. Prod.* **2012**, *40*, 225–231. [CrossRef]

26. Scheller, H.V.; Ulvskov, P. Hemicelluloses. *Annu. Rev. Plant Biol.* **2010**, *61*, 263–289. [CrossRef] [PubMed]

27. Galanakis, C.M.; Tornberg, E.; Gekas, V. Recovery and preservation of phenols from olive waste in ethanolic extracts. *J. Chem. Technol. Biotechnol.* **2010**, *85*, 1148–1155. [CrossRef]

28. Roselló-Soto, E.; Galanakis, C.M.; Brnčić, M.; Orlien, V.; Trujillo, F.J.; Mawson, R.; Knoerzer, K.; Tiwari, B.K.; Barba, F.J. Clean recovery of antioxidant compounds from plant foods, byproducts and algae assisted by ultrasounds processing. Modeling approaches to optimize processing conditions. *Trends Food Sci. Technol.* **2015**, *42*, 134–149. [CrossRef]

29. Stamatopoulos, K.; Katsoyannos, E.; Chatzilazarou, A. Antioxidant activity and thermal stability of oleuropein and related phenolic compounds of olive leaf extract after separation and concentration by salting-out-assisted cloud point extraction. *Antioxidants* **2014**, *3*, 229–244. [CrossRef]

30. Soler-Rivas, C.; Espín, J.C.; Wichers, H.J. Oleuropein and related compounds. *J. Sci. Food. Agric.* **2000**, *80*, 1013–1023. [CrossRef]

31. Bunzel, M.; Schendel, R.R. Determination of (Total) Phenolics and Antioxidant Capacity in Food and Ingredients. In *Food Analysis*, 5th ed.; Nielsen, S., Ed.; Springer Cham: New York, NY, USA, 2017; pp. 455–468.

32. Karki, B.; Lamsal, B.P.; Jung, S.; van Leeuwen, J. (Hans); Pometto, A.L.; Grewell, D.; Khanal, S.K. Enhancing protein and sugar release from defatted soy flakes using ultrasound technology. *J. Food Eng.* **2010**, *96*, 270–278. [CrossRef]

33. Jönsson, L.J.; Martín, C. Pretreatment of lignocellulose: Formation of inhibitory by-products and strategies for minimizing their effects. *Bioresour. Technol.* **2016**, *199*, 103–112. [CrossRef]

34. Lama-Muñoz, A.; Contreras, M.d.M.; Espínola, F.; Moya, M.; Romero, I.; Castro, E. 1. Optimization of oleuropein and luteolin-7-*O*-glucoside extraction from olive leaves by ultrasound-assisted technology. *Energies* **2019**, *12*, 2486. [CrossRef]

35. Salido, S.; Perez-Bonilla, M.; Adams, R.P.; Altarejos, J. Phenolic components and antioxidant activity of wood extracts from 10 main Spanish olive cultivars. *J. Agric. Food Chem.* **2015**, *63*, 6493–6500. [CrossRef] [PubMed]

36. Sari, Y.W.; Mulder, W.J.; Sanders, J.P.M.; Bruins, M.E. Towards plant protein refinery: Review on protein extraction using alkali and potential enzymatic assistance. *Biotechnol. J.* **2015**, *10*, 1138–1157. [CrossRef] [PubMed]

37. Sari, Y.W.; Syafitri, U.; Sanders, J.P.M.; Bruins, M.E. How biomass composition determines protein extractability. *Ind. Crop. Prod.* **2015**, *70*, 125–133. [CrossRef]

38. Zhang, C.; Sanders, J.P.M.; Bruins, M.E. Critical parameters in cost-effective alkaline extraction for high protein yield from leaves. *Biomass Bioenerg.* **2014**, *67*, 466–472. [CrossRef]

39. Lorenzo-Hernando, A.; Ruiz-Vegas, J.; Vega-Alegre, M.; Bolado-Rodríguez, S. Recovery of proteins from biomass grown in pig manure microalgae-based treatment plants by alkaline hydrolysis and acidic precipitation. *Bioresour. Technol.* **2019**, *273*, 599–607. [CrossRef]

40. Zhang, C.; Sanders, J.P.M; Xiao, T.T.; Bruins, M.E. How does alkali aid protein extraction in green tea leaf residue: A basis for integrated biorefinery of leaves. *PLoS ONE* **2015**, *10*, e0133046. [CrossRef]

41. Maayan, I.; Shaya, F.; Ratner, K.; Mani, Y.; Lavee, S.; Avidan, B.; Shahak, Y.; Ostersetzer-Biran, O. Photosynthetic activity during olive (*Olea europaea*) leaf development correlates with plastid biogenesis and Rubisco levels. *Physiol. Plant.* **2008**, *134*, 547–558. [CrossRef]

42. Chen, Y.; Stevens, M.A.; Zhu, Y.; Holmes, J.; Xu, H. Understanding of alkaline pretreatment parameters for corn stover enzymatic saccharification. *Biotechnol. Biofuels* **2013**, *6*, 8. [CrossRef]

43. Ponnusamy, V.K.; Nguyen, D.D.; Dharmaraja, J.; Shobana, S.; Banu, J.R.; Saratale, R.G.; Chang, S.W.; Kumar, G. A review on lignin structure, pretreatments, fermentation reactions and biorefinery potential. *Bioresour. Technol.* **2019**, *271*, 462–472. [CrossRef]

44. Kim, J.S.; Lee, Y.Y.; Kim, T.H. A review on alkaline pretreatment technology for bioconversion of lignocellulosic biomass. *Bioresour. Technol.* **2016**, *199*, 42–48. [CrossRef]

45. Tarasov, D.; Leitch, M.; Fatehi, P. Lignin–carbohydrate complexes: Properties, applications, analyses, and methods of extraction: A review. *Biotechnol. Biofuels* **2018**, *11*, 269. [CrossRef] [PubMed]

46. Rencoret, J.; Gutiérrez, A.; Castro, E.; del Río, J.C. Structural characteristics of lignin in pruning residues of olive tree (*Olea europaea* L.). *Holzforschung* **2019**, *73*, 25–34. [CrossRef]

47. McIntosh, S.; Vancov, T. Optimisation of dilute alkaline pretreatment for enzymatic saccharification of wheat straw. *Biomass Bioenerg.* **2011**, *35*, 3094–3103. [CrossRef]

48. Rommi, K.; Niemi, P.; Kemppainen, K.; Kruus, K. Impact of thermochemical pre-treatment and carbohydrate and protein hydrolyzing enzyme treatment on fractionation of protein and lignin from brewer's spent grain. *J. Cereal Sci.* **2018**, *79*, 168–173. [CrossRef]

49. Arzamasov, A.A.; van Sinderen, D.; Rodionov, D.A. Comparative genomics reveals the regulatory complexity of bifidobacterial arabinose and arabino-oligosaccharide utilization. *Front. Microbiol.* **2018**, *9*, 776. [CrossRef]

Article

Biocompounds Content Prediction in Ecuadorian Fruits Using a Mathematical Model

Wilma Llerena [1,2], Iván Samaniego [3], Ignacio Angós [1], Beatriz Brito [3], Bladimir Ortiz [1] and Wilman Carrillo [4,*]

[1] Facultad de Ciencia e Ingeniería en Alimentos y Biotecnología, Universidad Técnica de Ambato (UTA), Av. Los Chasquis y Río Payamino, 180103 Ambato, Ecuador
[2] Facultad de Ciencias Pecuarias, Ingeniería en Alimentos, Universidad Técnica Estatal de Quevedo, Km 7 1/2 vía Quevedo-El Empalme, 120313 Los Ríos, Ecuador
[3] Departamento de Nutrición y Calidad, Instituto Nacional de Investigaciones Agropecuarias (INIAP), Panamericana Sur Km. 1, 170516 Mejía, Ecuador
[4] Departamento de Investigación, Universidad Técnica de Babahoyo, Av. Universitaria Km 2 1/2 Av. Montalvo, 120301 Babahoyo, Ecuador
* Correspondence: wcarrillo@utb.edu.ec; Tel.: +593-980288016

Received: 22 June 2019; Accepted: 23 July 2019; Published: 25 July 2019

Abstract: Anthocyanins, carotenoids and polyphenols are biomolecules that give the characteristic color to fruits. Carotenoids relate to yellow, orange and red colors whereas anthocyanins and polyphenols mainly relate to purple and red colors. Presently, standard determination of antioxidants is carried out using relatively complex methods and techniques. The aim of this study was to develop a mathematical prediction model to relate the internal color parameters of the Amazonic fruits araza (*Eugenia stipitata* Mc Vaugh), Andean fruit blackberry (*Rubus glaucus* Benth), Andean blueberry (*Vaccinium floribundum* Kunth), goldenberry (*Physalis peruviana* L.), naranjilla (*Solanum quitoense* Lam.), and tamarillo (*Solanum betaceum* Cav.) to their respective anthocyanins, carotenoids and polyphenols contents. The mathematical model was effective in predicting the total anthocyanins content (TAC), the total carotenoids content (TCC) and finally the total phenolic content (TPC) of fruits assayed. Andean blueberry presented a TPC with an experimental value of 7254.62 (mg GAE/100 g sample) with respect to a TPC prediction value of 7315.73 (mg GAE/100 g sample). Andean blackberry presented a TAC with an experimental value of 1416.69 (mg chloride cyanidin 3-glucoside/100 g) with respect to a prediction TAC value of 1413 (mg chloride cyanidin 3-glucoside/100 g).

Keywords: chemometrics; mathematical model; metaheuristic techniques; color; araza; blackberry; Andean blueberry; naranjilla; tamarillo; goldenberry

---

1. Introduction

Due to its geographical location, Ecuador is a diverse country in terms of climate and fruit production. Fruit consumption is clearly associated with health benefits such as enhancing the immunologic system, reduction of cellular oxidative damage and protection against cancer development [1]. These properties are attributed to the presence of phytochemicals and nutrients with antioxidant properties [2–4]. Antioxidants can have chemoprotective effects, which include prevention of cardiac diseases, antidiabetic activity and vasoprotective properties [5–7].

Antioxidants can be classified in four phytochemicals main groups: phenolic compounds, including anthocyanins, terpene substances, including carotenoids, sulphur compounds, and finally, nitrogen compounds alkaloids. Amongst them, the first three groups are the most important as bioactive constituents in fruits are responsible for skin and pulp color [8].

Polyphenols are responsible for the red, blue and purple colors in many fruits and legumes [9,10]. In this group, anthocyanin is a group of water-soluble pigments, composed by a molecule of anthocyanidin, also named aglycone, linked to a sugar with a β-glucosides bond [1]. Its color and stability depend on several factors such as chemical structure, pH and temperature [5]. Anthocyanins are associated with flavylium cation, which produces red color at low pH $\leq$ 1.0 [1]. Concerning carotenoids, these compounds are common natural pigments. Around 600 carotenoids have been described in the literature, β-carotene being the most representative. Carotenoids pigments are responsible for red, orange, and yellow hues of plant leaves, fruits, and flowers, as well as the colors of some birds, insects, fish, and crustaceans. Plants, bacteria, fungi, and algae can synthesize carotenoids. However, animals and humans incorporate carotenoids through their diet. Some carotenoids can be used as a source of vitamin A [11].

During ripening, a series of biochemical and physiological processes occur, producing changes in the texture, flavor and color of the fruits [12,13]. Color changes are evident during fruit development and ripening and keep on going after harvesting. Orange color becomes evident in β-carotene rich fruits as araza, naranjilla, tamarillo and goldenberry fruits, when the degreening process occurs. During the maturation stage, carotenoids build up and at the same time, chlorophylls start a degradation process to a pheophytin form [13,14]. Fruits of deep red and blue colors such as blackberry and Andean blueberry show a superficial color change due to an accumulation of anthocyanins associated with changes in the concentration of sugars and organic acids [15].

Determination of fruit properties is of utmost importance in the food industry, color being the most important of visual attributes. Consumers use color as a fast index of quality of a fruit. Color is associated with its taste, freshness and nutritional value. Moreover, color is related to fruit shelf life: a reduction in its characteristic traits can be easily associated with decay [16–18].

Color and antioxidant content can be quantified based on the characteristic radiation wavelength absorption of each pigment in the visible region of the electromagnetic spectrum [19,20]. The classical spectrophotometric methods for determination of light absorption are carried out with equipment and specialized procedures, not easily affordable for small-medium scale enterprises. Nevertheless, quality can be related to color in an easier way, using portable spectrophotometric methods associated to a wide gamut color space L* a* b* coordinates [21–23]. From these measurements, non-destructive chemometric prediction methods can be created using deterministic and stochastic mathematical models. These models are fast, precise and easy enough to be used in routine food quality control [24–26].

The aim of the present work was to determine the total content of anthocyanins, carotenoids and polyphenols, of six representative Andean and tropical fruits from Ecuador using the UV-visible spectrophotometry and to develop a mathematical tool to predict the nutritional value based on the measurement of the internal color, as a cheap and fast alternative quality method of analysis. Antioxidant activity was also determined.

2. Materials and Methods

2.1. Raw Material

Two tropical fruits, and four Andean fruits were chosen in this research. Araza (*Eugenia stipitata* Mc Vaugh) is a fruit from the genus *Mirtaceae* cultivated all around the Amazonic basin. Two clones (INIAP 001 and INIAP 003) from the Orellana province in Ecuador were selected for their special aptitude to be industrially processed due to their unique flavor, strong sourness and short shelf life.

Blackberry (*Rubus glaucus* Benth) is a fruit native from the high lands of the intertropical region [27]. Cultivar 'INIAP Andimora 2013' is an improved clone cultivated in the Tungurahua province. This clone is thornless, has a high yield, high fruit quality and improved resistance to the main diseases affecting this plant.

Mortiño or Andean blueberry (*Vaccinium floribundum* Kunth) is a small perennial bush growing wild in the highlands in the Andes. Fruits are spherical berries of dark blue color, traditionally harvested in the Ecuadorian provinces of Bolivar, Cotopaxi, and Pichincha.

Naranjilla (*Solanum quitoense* Lam.) belongs to the Solanaceae family, native of the Andean medium ranges of Ecuador, Colombia and Central America [28]. The Instituto Nacional de Investigaciones Agropecuarias (INIAP) from Ecuador, through the National Program of Fruticultura, has generated technologies that allow generating resistant materials and practices of integrated agronomic management. One of these materials is the juicy naranjilla INIAP Quitoense 2009, which comes from a selection of the variety 'Baeza' years 2005–2007, and presents better characteristics in terms of vigor, yield capacity, productivity and physicochemical quality of fruits.

Tamarillo (*Solanum betaceum* Cav.) is a Solanaceae with a medium-sized fruit, oval berry, with a juicy bitter-sweet pulp. Ecotype 'Anaranjado Gigante' comes originally from the Tungurahua province and is mainly featured by its light orange pulp color [29].

Goldenberry (*Physalis peruviana* L.) is an annual or short-lived culture belonging to the genus *Solanaceae*, which is mostly grown for its ecotype 'Golden Kenyan' in the Tungurahua province of Ecuador. The fruit, usually commercialized with its distinctive protector calyx, is small, round and yellow with a medium sourness flavor [30].

2.2. Sample Preparation

Fifteen kilograms of each fruit was recollected from trees to ensure the heterogeneity. The fruits were washed with drinking water to reduce the microbial load, dirt and organic matter. They were then separated in portions of 1 kg to obtain 15 samples for each fruit assayed. Then, the maturity index was determined, and the fruit was homogenized, screened and stored in high-barrier plastic bags with hermetic seals at −18 °C, out of oxygen and light. The determination of internal color, anthocyanins, and total polyphenols was carried out taking into consideration 15 samples per fruit, in triplicate (n = 45 for each fruit). The total carotenoid content determination was carried out in duplicate (n = 30 for each fruit). The total flavonoids content was determinate only in three fruits (araza, naranjilla and tamarillo) and was made for triplicate (n = 45). All methods were validated using CV Horwitz—15 samples for fruits were used and duplicate and triplicate measurements were taken. CV Horwitz were considered significant with values ≤16.0%.

2.3. Physicochemical Analysis

2.3.1. Texture

Fruit firmness was determined by puncture with two penetrometers Gullimex (Borne, Netherlands), using different probe diameters. The results were expressed in Newtons (N). Araza, naranjilla and tamarillo were tested using a FT327 model with a cylindrical probe of 6-mm diameter. Goldenberry, blackberry and Andean blueberry were tested with a FT011 model with a cylindrical probe of 2-mm diameter.

2.3.2. Maturity Index

The maturity index (MI) was calculated based on the titratable acidity and the content of total soluble solids. Titratable acidity was obtained by acid-base neutralization according to Ecuadorian standards. The result was expressed in terms of percent of malic acid for araza [31] and blackberry [12] and citric acid for Andean blueberry [32], tamarillo [33], naranjilla [34], and goldenberry [35].

Total soluble solid (TSS) content was directly measured in the pulp of each sample with a digital handheld refractometer Atago, model PAL, 0–53 °Brix (Tokio, Japan). The results were expressed as g of sucrose per 100 g of sample (°Brix).

The MI was calculated with Equation (1), Tehranifar et al. (2010) [36].

$$MI = TSS/TA \tag{1}$$

where,

MI: maturity index (dimensionless)
TSS: total soluble solid content (°Brix)
TA: titrable acidity (g/100 g)

2.3.3. Internal Color

The internal color was determined with a handheld colorimeter ColorTec-PCM (ColorTec, Clinton, NJ, USA) with a measurement angle of 10°, D65 illuminant and 8 mm aperture. Chromatic properties of the fruit pulps were expressed in the CIE (Commission Internationale de l'Eclairage) L*a*b* color space in terms of coordinates L* luminosity, a* red/green and b* blue/yellow. From each fruit, 500 mL of pulp was extracted and homogenized. Samples of 30 mL were carefully poured in a Petri dish avoiding lumps or bubbles. The Petri dish was placed over a white surface and divided into four equal areas. Duplicate measurements were done in each quarter and center of the plate.

2.4. Total Anthocyanins Content (TAC)

TAC was determined using the differential pH method used by Rapisarda et al. (2000) [37]. The extraction was done with a magnetic stirrer for 60 min, taking 0.25 g of freeze dried sample, adding 10 mL of buffer solution at pH 1.0 (potassium chloride 0.2 N and hydrochloric acid 0.2 N) and a buffer solution at pH 4.5 (sodium acetate 1 M, hydrochloric acid 1 N). After centrifugation of the extract at 5000 rpm, 1 mL of the solution was diluted with the buffer solution to 10^{-3} for buffer pH 1.0 and 10^{-1} for buffer pH 4.5. The absorbance was measured in the supernatant and buffer solutions at 510 nm and 700 nm with a UV-VIS spectrophotometer Shimadzu, model 2200 (Shimaszu, Kioto, Japan). Results were expressed as mg of cyanidin-3-glucoside chloride/100 g of the dry weight sample (DW). TAC was calculated based on the following equation:

$$TAC = A \times MW \times DF \times 100/\varepsilon \times W$$

where A is the absorbance, MW molecular weight of cyanidin-3-glucoside chloride ($C_{21}H_{21}ClO_{11}$, 484.84 g/moL), DF is dilution factor, ε molar absorptivity (34,300), W = sample weight (g).

2.5. Total Carotenoids Content (TCC)

TCC was measured in the absence of light and oxygen, using 0.6–1.0 g of the freeze-dried sample. The extraction was done using 50 mL of a solvent mixture composed by hexane 50%, ethanol 25%, acetone 25% (*v/v/v*), 0.1% of butylated hydroxytoluene (BHT) (p/v) and 5 g of calcium chloride (p/v). These elements were added gradually. The mixture was mixed for 20 min in a refrigerated water bath at 4 °C. The phase separation was achieved by adding 15 mL of distilled water for 10 min. The extract was filtered and transferred to a separating funnel. The organic phase was transferred to a volumetric flask. Hexane was added to reach 50 mL. The determination of total carotenoids content was made using the UV-VIS spectrophotometer Shimadzu, model 2200 (Kioto, Japan) at 450 nm, using the method of Leong & Oey (2012) [38]. The results were expressed in terms of μg of β-carotene/μg of dry weight basis (DW). TCC was calculated based on the following equation:

$$TCC = A \times VT \times 10^4/2592 \times W$$

where A is the absorbance at 450 nm, VT is the volume total, 2592 is the coefficient of extinction molar of β-carotene in hexane and W is the weight of the sample and 10^4 constant of conversion of units of μg/g.

2.6. Total Polyphenols Content (TPC)

TPC was determined following a method by Georgé et al. (2005) [39]. The extraction was made using a solution of 70% acetone (*v/v*) under magnetic stirring for 45 min, using 0.3–1.0 g of freeze-dried samples. The mixture was centrifuged for 10 min at 3500 rpm and the supernatant raw extract was recovered in Eppendorf vials. TPC A non-soluble fraction and the soluble compound TPC B fraction were evaluated from the raw extract. For the A fraction, a series of solutions of 25, 50 and 75 µL of the raw extract in 500 mL of pure methanol were prepared. The separation of soluble compounds was done by solid phase extraction (SPE) using C18 OASIS cartridges (Waters Corp.; Milford, MA, USA) previously conditioned following the fabricant instructions. 500 µL of raw extract were diluted in 3.5 mL of distilled water and a 2 mL aliquot of this solution was injected into the OASIS cartridge.

The quantification of A and B fractions was done using the Folin-Ciocalteu method with a UV-VIS spectrophotometer Shimadzu 2200 (Kioto, Japan) at 760 nm and both fractions were used (Abs_B-Abs_A) to corrected interferences. Gallic acid was used as standard. The curve was stablished ($y = 0.0011x + 0.0529$, $R^2 = 0.9997$). The total polyphenol content was expressed in terms of mg of gallic acid equivalents GAE/100 g of the DW sample.

2.7. Mathematical Modeling

Mathematical modeling was carried out using chemometric techniques of pattern recognition by object representation in a multidimensional space towards a reduced dimensionality space [25]. The correlation matrix between dependent and independent variables was determined using a multidimensional regression and the mean square method (Equations (2)–(4)). Finally, a results matrix was obtained (Table 1).

$$A \sum_{K=1}^{N} x^4 k + B \sum_{K=1}^{N} x^3 k + C \sum_{K=1}^{N} x^2 k = \sum_{K=1}^{N} x^2 k Y k \tag{2}$$

$$A \sum_{K=1}^{N} x^3 k + B \sum_{K=1}^{N} x^2 k + C \sum_{K=1}^{N} x k = \sum_{K=1}^{N} x k Y k \tag{3}$$

$$A \sum_{K=1}^{N} x^2 k + B \sum_{K=1}^{N} x k + C N = \sum_{K=1}^{N} Y k \tag{4}$$

In Table 1, color coordinates L*, a* and b* were the independent variables. TAC, TCC and TPC acted as the dependent variables for the n samples.

Table 1. Multivariate components matrix.

Biocompounds Content	Luminosity (L*)	Coordinate Red/Green (a*)	Coordinate Yellow/Blue (b*)
Y_1	X_{11}	X_{12}	X_{13}
Y_2	X_{21}	X_{22}	X_{23}
Y_n	X_{n1}	X_{n2}	X_{n3}
Mean	X_1	X_2	X_3

The robustness analysis of the mathematical prediction model was evaluated using the coefficient of determination. The data homogeneity was evaluated using the experimental residual analysis vs the predicted values. Only samples with standard deviations lower than 2σ were used [17]. The partial coefficients of the regression model were tested for signification (Equation (5)) to verify if parameters L*, a* and b* added value to the prediction model.

$$t = \frac{bi}{\sqrt{Pij * S^2}} \tag{5}$$

where,

bi: partial regression coefficient
Pij: i-row and j-column of the reverse square sum and cross-product matrix
S^2: estimator for the variance of the standard deviations and residues

The magnitude of the correlations between the variables was evaluated with the values of the determination coefficient (R^2). The most influential variables in the model and the data with the better adjustments were used for a new multiple regression analysis, establishing the final mathematical model (Equation (6)).

$$Y_c = b_0 + b_1 X_1 + b_2 X_2 + b_3 X_3 \tag{6}$$

where,

Y_c: the content of total antioxidant: anthocyanins, carotenoids and polyphenols
b_1, b_2, b_3: regression coefficients
X_1, X_2, X_3: color parameters L*, a* and b*

2.8. ABTS Assay

The extracts of araza (*Eugenia stipitata* McVaugh), tamarillo (*Solanum betaceum* Cav.) and naranjilla (*Solanum quitoense* Lam.) were used to evaluate their antioxidant activity using the ABTS method described by Piñuel et al. (2019). Trolox was used as the reference standard (0–800 μmoL Trolox/L). The curve was established (y = 0.0007x + 0.0671, R^2 = 0.999). The results obtained were expressed as μmol Trolox Equivalents TE/ g sample. All assays were made in triplicate [40].

2.9. DPPH Assay

The extracts of araza (*Eugenia stipitata* McVaugh), tamarillo (*Solanum betaceum* Cav.) and naranjilla (*Solanum quitoense* Lam.) were used to evaluate the antioxidant activity using the DPPH method described by Piñuel et al. (2019). Trolox was used as the reference standard (50–500 μmol Trolox/L) and the curve was established (y = 0.0013x + 0.007, R^2 = 0.999). The results obtained were expressed as μmol TE/g sample. All assays were made in triplicate [40].

2.10. Statistical Analysis

The results obtained in this study were presented as means ± standard deviation (SD). Differences between group values were determined using the one-way ANOVA analysis, followed by the Tukey's test. All tests were considered with statistical differences at $p < 0.05$ and $p < 0.01$ using the software Graph Pad Prism 4. Moreover, the data obtained with the mathematical model were processed with the Statistica 10.0 software to obtain the analysis graphs.

3. Results and Discussion

3.1. Maturity Index

Climacteric fruits as araza, naranjilla and tamarillo can reach full ripening after harvesting. These fruits were harvested at physiological maturity. The MI is a parameter commonly used to establish the commercial maturity of these products [12]. As shown in Table 2, naranjilla and tamarillo presented an MI over 2.5 and 4.5 respectively. Araza is extremely perishable. The fruit used in this work, presented an MI between 1.15 and 2.05, corresponding to a partially mature green-yellow and over ripened full yellow color [41].

Non climacteric fruits, blackberry, Andean blueberry and goldenberry, were harvested at the final stage of edible maturity: stage 6 for blackberry [42] and Andean blueberry [32] and stage 5 for goldenberry. MI reached 9.64, very close to the value MI of 10.45 reported by Fischer et al. (2011) [35].

Changes during the postharvest period can modify firmness and color of fruits [13]. As can be seen in Table 2, naranjilla and tamarillo continued their ripening during storage and transportation, something related to their high variability observed in firmness of 36.52% and 29.69% respectively. Blackberry and araza suffered a firmness loss due to mechanical damages during transport. This data is not shown in the text. High water content and soft flesh make araza very susceptible to damage. Moreover, its exceptionally high respiration rate can change dramatically its color and firmness in a period as short as 72 h after being harvested [43]. In the same way, goldenberry presented a high variability in its firmness—14.72% at maturity—that could be attributed to being harvested with different MIs. This did not happen in the case of Andean blueberry.

Table 2. Physical and chemical parameters related to the maturity stage of the fruits.

Sample	Titrable Acidity * (%)	Total Soluble Solids * (°Brix)	Maturity Index *	Firmness * (N)
Araza	2.40 ± 0.02	3.83 ± 1.19	1.60 ± 0.45	14.42 ± 5.00
Blackberry	2.81 ± 0.07	12.69 ± 0.43	4.51 ± 0.06	3.22 ± 0.49
Andean blueberry	0.96 ± 0.05	11.81 ± 0.26	12.39 ± 0.72	0.69 ± 0.01
Naranjilla	2.58 ± 0.15	9.55 ± 0.43	3.72 ± 0.32	48.05 ± 17.55
Tamarillo	2.09 ± 0.05	12.43 ± 0.94	5.94 ± 0.37	55.80 ± 16.57
Goldenberry	1.42 ± 0.02	13.73 ± 1.50	9.64 ± 0.62	2.65 ± 0.39

* Fresh basis.

3.2. Internal Color

The relationship between color and ripeness is due to the pigment accumulation and variation of the sugar and organic acid in fruits [44]. In Table 3 and Figure 1, it can be observed the internal color of the fruits was represented in the CIE L*a*b* color space. Araza and naranjilla pulp presented a clear trend towards green and yellow colors with an intermediate luminosity L* 49.63 and 40.10, respectively. Tamarillo and goldenberry presented a more yellow-reddish color and more differences in luminosity L*, 51.71 and 35.70, respectively. Blackberry and Andean blueberry presented the lowest values for luminosity L* 10.68 and 20.80, respectively, according to Hue et al., 2014 [17].

Table 3. CIE L*a*b* color coordinates in several tropical and Andean fruits.

Fruit	Color Coordinates							
	Lightness		Red-Green		Blue-Yellow		Chroma	Hue
	L*	CV (%)	a*	CV (%)	b*	CV (%)	C*	°H
Araza	49.63 ± 2.94	5.91	−0.89 ± 0.40	45.32	22.73 ± 2.84	12.48	92.31 ± 1.20	22.75 ± 2.83
Blackberry	10.68 ± 2.23	20.88	13.80 ± 3.90	28.26	4.95 ± 1.70	34.34	19.86 ± 4.94	14.71 ± 27.85
Andean blueberry	20.80 ± 1.50	7.21	3.52 ± 1.10	31.25	3.16 ± 0.95	30.06	42.28 ± 13.38	4.86 ± 0.87
Naranjilla	40.10 ± 1.92	4.79	−4.25 ± 0.60	14.03	22.04 ± 2.62	11.88	100.93 ± 1.05	22.45 ± 2.65
Tamarillo	51.75 ± 2.93	5.67	9.06 ± 0.71	7.84	32.68 ± 2.83	8.65	74.45 ± 1.41	33.92 ± 2.80
Goldenberry	35.70 ± 2.46	6.88	7.10 ± 0.51	7.25	25.39 ± 3.55	13.99	74.11 ± 2.48	26.39 ± 3.41

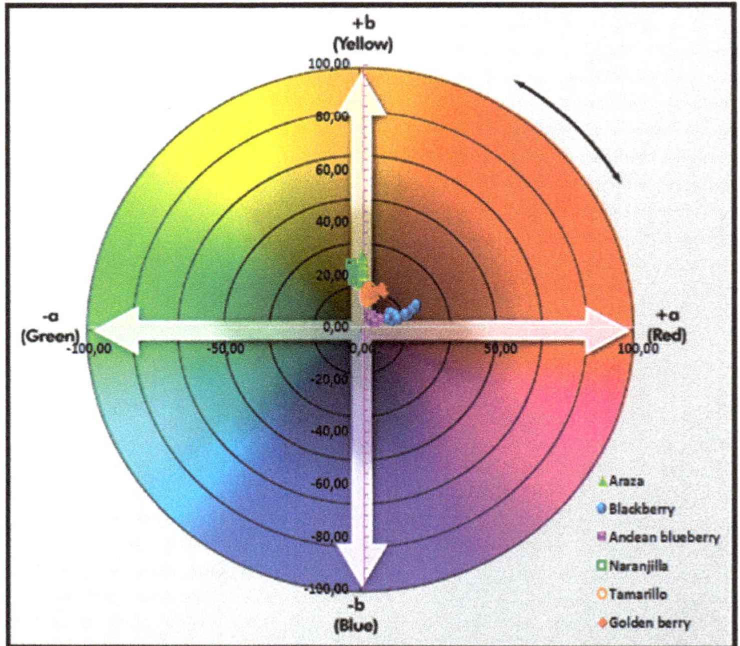

Figure 1. Chromatic representation of the CIE L*a*b* color coordinates of the tropical and Andean fruits.

The rupture of the chlorophyll and accumulation of carotenoids is a phenomenon that occurs in araza, naranjilla, tamarillo and goldenberry. This phenomenon determines its color at ripeness, and turns the color of these fruits from green to yellow. β-carotene is the center of this photosynthetic reaction, its concentration in fruits being related to variety, harvest time and other factors like soil, climate, cultural practices, etc. [13]. In the case of blackberry and Andean blueberry, the color was dominated by red and blue, due to the characteristic accumulation of anthocyanins during ripening of these fruits [45].

All fruits evaluated presented polyphenols in their composition, Andean blueberry and blackberry being the fruits with the highest concentration of this antioxidant.

3.3. Development of the Mathematical Prediction Models

Correlation matrices were constructed using color coordinates L*, a* and b* as independent variables and TAC, TPC and TCC as dependent variables. Anthocyanin and polyphenol content models were developed for blackberry and Andean blueberry, whereas carotenoid and polyphenol content models were developed for araza, naranjilla, tamarillo and goldenberry.

Mathematical equations were evaluated considering two aspects, significance of the coefficients of the independent variables of the model and the global determination coefficient of the model. The mathematical equations were evaluated considering two aspects: significance of the coefficients of the independent variables of the model and the global determination coefficient of the model. The whole set of coordinates was used in order to elaborate the mathematical prediction models for color, as there were no significant differences for the partial coefficients of each coordinate L*, b*, and b* in terms of anthocyanin, carotenoid and total polyphenol contents.

The first modelling approach with the whole data resulted in very low determination coefficients for the TAC and TCC of all fruits. Based on the analysis of residuals between the theoretical and experimental data, the results obtained were consistent series of data falling inside a confidence interval of two standard deviations from the mean.

The selected data was submitted to a new multivariate analysis. The adjusted prediction equations for each fruit (Table 4; Table 5) were determined. The TAC prediction models had determination coefficients R^2 of 0.84 and 0.98, whereas the TPC presented R^2 between 0.81 and 0.89.

From the equations in Table 4, a prediction of bioactive compounds for each fruit was calculated in terms of TAC, TCC, and TPC. This result was compared to the experimental and bibliographic data in Table 5. The developed method of analysis was validated using the Horwitz variation coefficient (CV Horwitz). This parameter relates the concentration of analyte with the coefficient of variation of the experimental data. CV Horwitz was proposed as a reference value to evaluate the inter-laboratory tests performance. It has been accepted by the EU, IUPAC and CODEX to validate analytical methods. The value of CV Horwitz should be ≤16.0% [46].

3.4. Total Anthocyanin Content (TAC)

Anthocyanins were reported in mg/100 g accepting an error of 5–8%. As can be seen in Table 5, blackberry presented a TAC of 1416.68 ± 158.71 mg/100 g, DW, very close to the 637–3000 mg/100 g interval reported for different blackberry cultivars and harvesting conditions. The prediction model for blackberry showed a value of 1413 mg/100 g very close to the experimental data and with a coefficient of variation estimated (CV_E) of 11.20.

Andean blueberry showed higher values of TAC of 2682.30 ± 602.92 mg/100 g close to data reported by Vasco et al. (2009) [47] around 3832.95 mg/100 g DW. The prediction model showed a good approximation to experimental data, offering a value of 2761.24 mg/100 g. In this case, the coefficient of variation predicted (CV_P) of 2.66 was lower to the experimental error of CV_E of 5.74.

Blackberry and Andean blueberry showed red and violet dominance, correlated significantly with TAC, obtaining R^2 values of 0.82 y 0.81, respectively.

Anthocyanins are part of a group of bioactive compounds present in the pulp of Andean blueberry and blackberry, and responsible for their characteristic red-blue color. In the case of araza, naranjilla, tamarillo and goldenberry, the presence of these chemical compounds was not identified, or their concentrations were under the detection limit of the spectrophotometric method employed. However, significant amounts of carotenoids were found in these fruits with characteristic yellow-orange color, as shown in Section 3.6.

3.5. Total Polyphenol Content

Polyphenols were reported in mg/100 g accepting an error of 5–8%. Polyphenols showed a significant correlation with the color of the six fruits studied, and especially those with red and violet colors, obtaining coefficients of 0.88 for araza, 0.81 for blackberry, 0.82 for mortiño, 0.84 for naranjilla, 0.75 for tamarillo and 0.88 in goldenberry.

As can be seen in Table 6, Andean blueberry showed the highest TPC, 7254.62 ± 1209.17 mg/100 g; CV_E 10.86%. These results are in accordance with the ones reported by several authors working with Ecuadorian Andean blueberry of 8104.52–9799.02 mg/100 g [48]. The prediction models showed a value of 7315.73 mg/100 g with a CV_P of 2.22%.

Blackberry showed the second highest TPC of 6352.28 ± 633.61 mg/100 g; CV_E 4.47% falling into the upper part of the interval 2340.24−6300 mg/100 g reported by several authors [49]. The prediction model showed a theoretical value for this fruit of 5995.62 mg/100 g, with a CV_P of 3.38% (Table 5).

Araza also showed a high TPC of 3507.79 ± 1430.36 mg/100 g with a CV_E 4.65% in accordance to the results reported by Laverde-Acurio, (2010) [50] for clone 003 (2477.72 mg/100 g). In this case, the predicted value for TPC (3256.33 mg/100 g) resulted in a CV_P of 0.98% (Table 5).

Tamarillo from ecotype Orange Giant showed a TPC of 1062.77 ± 57.87 mg/100 g; with a CV_E 10.26%. Torres (2006) [51] showed a value for this fruit of 654.20 mg/100 g DW. The predicted value from equations obtained resulted in 1055.45 ± 14.80 mg/100 g with a CV_P 0.72% (Table 5).

Naranjilla cv. INIAP quitoense 2009 showed a TPC of 897.58 ± 227.77 mg/100 g and a CV_E of 6.32%, above the 510.72–699.79 mg/100 g interval reported by several authors for several agroclimatic conditions [51]. The predicted value for TPC in naranjilla was 730.84 mg/100 g, with a CV_P of 2.02% (Table 5).

Experimental TPC obtained for goldenberry was 259.93 ± 42.74 mg/100 g with a CV_E 11.61%, whereas the predicted TPC resulted in 233.68 mg/100 g, with a CVP of 3.24%. Both results are lower than bibliographic data reported by Cerón et al. (2010) in goldenberry grown in Colombia [52].

Experimental errors for Andean blueberry of 10.86%, tamarillo with 10.26% and goldenberry with 11.61% were slightly higher than the upper limit imposed by Horwitz criterium for this range of concentrations, 5–8%.

3.6. Total Carotenoid Content

Carotenoid concentration was measured in µg/g, accepting an error of <16%. Table 5 shows the TCC results: tamarillo showed a dry basis TCC of 123.18 ± 16.61 µg/g with a CV_E 4.65%, the highest concentration of TCC among all fruit studied and very similar to data reported by Mertz et al. (2009) [45] for the same cultivar and origin of 117.37 µg/g. The prediction model showed a TCC of 133.67 µg/g with a CV_P of 0.98%. Both CV were well above the Horwitz limit for this range of concentrations around 16%.

Goldenberry showed a TCC of 65.21 ± 8.31 µg/g with a CV_E 10.84% very close to data reported by Ramadan, (2011) who obtained a TCC of 85.38 µg/g on a dry basis for the ecotype Golden Kenyan. The prediction value for this fruit of 64.93 µg/g showed an estimated error of 6.49% also lower than the Horwitz limit [30].

In araza, an experimental TCC of 62.85 ± 3.36 µg/g with a CV_E 1.92% and a predicted value of 61.96 µg/g with a CV_P 1.09% were found. These results are close to 55.32 µg/g, DW, in accordance with the ones reported by Laverde-Acurio (2010), in clone 003 [50].

Finally, naranjilla (var. INIAP quitoense 2009) showed the lowest concentration of experimental TCC, 57.93 ± 4.28 µg/g with a value of CV_E 6.05% and a predicted a value of 58.42 µg/g, with a value of CV_P 4.50%, relatively close to data reported by Acosta et al. (2009) [34] in naranjilla from Costa Rica of 76.59 µg/g, DW.

The results obtained for araza, naranjilla, tamarillo and goldenberry confirmed that TCC found in these fruits are strongly associated with color parameters L*, a* and b* and thus being responsible for the yellow-orange color of the pulp of these fruits, obtaining coefficients R^2 ranged from 0.85 to 0.98 for the latter fruits. Also, it is worth noting that all experimental and predicted values for TCC were within the threshold reported by Horwitz for the concentration range of µg/g with a value of CV Horwitz 16%, as commented before.

Table 4. Mathematical prediction models for total anthocyanin content (TAC), total carotenoid content (TCC) and total polyphenol content (TPC) of six tropical and Andean fruits from Ecuador.

Fruits	$TAC = b_0 + b_1(L^*) + b_2(a^*) + b_3(b^*)$					$TCC = b_0 + b_1(L^*) + b_2(a^*) + b_3(b^*)$					$TPC = b_0 + b_1(L^*) + b_2(a^*) + b_3(b^*)$				
	b_0	b_1	b_2	b_3	R^2	b_0	b_1	b_2	b_3	R^2	b_0	b_1	b_2	b_3	R^2
Blackberry	1644.47	−13.53	−12.56	18.42	0.82	N. D	N. D	N. D	N. D	N. D	8375.00	−41.15	−132.30	18.79	0.82
Andean blueberry	−3206.38	233.19	96.97	264.9	0.81	N. D	N. D	N. D	N. D	N. D	7095.72	33.44	143.50	−308.63	0.82
Naranjilla	N, D	N. D	N. D	N. D	N. D	129.24	−1.04	1.47	−0.98	0.84	1151.18	−15.69	−74.94	−5.58	0.84
Tamarillo	N. D	N. D	N. D	N. D	N. D	79.29	−1.12	3.93	3.35	0.98	1247.55	−5.53	53.00	−11.51	0.75
Goldenberry	N. D	N. D	N. D	N. D	N. D	−24.31	2.19	2.41	−0.23	0.85	−142.76	9.85	10.09	−1.91	0.88
Araza	N. D	N. D	N. D	N. D	N. D	83.52	−0.84	−5.88	0.7	0.93	−4974.64	381.95	932.80	−450.89	0.88

TAC: total anthocyanin content; TCC: total carotenoid content; TPC: total polyphenol content and N.D: non- detectable.

Table 5. Experimental and predicted values for total anthocyanin content (TAC), total carotenoid content (TCC) and total phenolic content (TPC) of six tropical and Andean fruits from Ecuador and a comparison with data reported in literature.

Fruit	Antioxidant Compounds												
	* (TAC)				* (TCC)					* (TPC)			
	(mg cyanidin 3-glucoside chloride/100 g)				(µg β-carotene/g)					(mg GAE/100 g)			
	Experimental		Predicted		Bibliographic	Experimental		Predicted		Bibliographic	Experimental		Predicted		Bibliographic
	Mean	CV_E	Mean	CV_P		Mean	CV_E	Mean	CV_P		Mean	CV_E	Mean	CV_P	
Blackberry	1416.69	11.20	1413.00	0.85	637–3000 [27,46]	N. D					6352.28	4.77	5995.62	3.38	2340.24–6300 [27,46,53]
Andean blueberry	2682.30	2.66	2761.24	5.74	3832.95 [48]	N. D					7254.62	10.86	7315.73	2.22	8104.52–9799.02 [48,54]
Naranjilla	N. D					57.93	6.05	58.42	2.36	76.59 [55]	897.58	6.32	912.50	2.02	510.72–699.79 [49,50,53]
Tamarillo	N. D					123.18	4.65	133.67	1.95	117.37 [46]	1062.77	10.26	1055.45	0.72	654,20 [52]
Goldenberry	N. D					65.21	10.84	64.93	3.10	85.30 [30]	259.93	11.61	233.68	3.24	215.60 [34]
Araza	N. D					62.85	1.92	61.96	1.09	55.32 [51]	3507.79	13.97	3256.33	6.94	[51]

* Results are reported as dry basis ± standard deviation (n = 45) for each fruit. N.D: non-detectable; GAE: gallic acid equivalents; CV_E: coefficient of variation estimated; CV_P: coefficient of variation predicted.

3.7. Antioxidant Activity Using ABTS and DPPH Methods

The extracts obtained from araza ((*Eugenia stipitata* McVaugh), tamarillo (*Solanum betaceum* Cav.) and naranjilla (*Solanum quitoense* Lam.) were used to evaluate their antioxidant activity using the ABTS and DPPH methods. Araza fruit presented a higher value of ABTS and DPPH with 758.22 μmoL TE/g sample and 392.10 μmoL TE/g, sample respectively. Tamarillo presented an ABTS value of 161.04 μmoL TE/g sample and 47.82. μmoL TE/g sample (Table 6). The value of ABTS and DPPH between fruits presented statistical differences at $p < 0.05$. Naranjilla fruit presented a low value of antioxidant activity with both methods. Espin et al. (2016) described the antioxidant activity of four varieties of tamarilllo fruits (*S. betaceum* Cav.) using the ABTS, FRAP and ORAC methods. ABTS analysis presented a value between 22 and 89 μmoL TE/g of the sample, lower than the one reported here [54]. It can be explained by the fact that the fruits used in our study were obtained from different cultivars and locations in Ecuador. Araza fruit has a high antioxidant activity with both methods and can be related to its high TPC content of 3507.79 mg GAE/100 g DW because it does not register TAC and its content of TCC was low; i.e., the experimental value of 62.81 μg β-carotene/g and 61.96 μg β-carotene/g predicted by the mathematical model.

Table 6. Antioxidant activity of araza, tamarillo and naranjilla fruits using the ABTS and DPPH methods.

Fruit	Antioxidant Activity (μmoL TE/g Sample)	
	ABTS Assay	**DPPH Assay**
Araza	758.22 ± 5.01 [a]	392.10 ± 9.67 [a]
Tamarillo	161.04 ± 8.48 [b]	47.82 ± 2.94 [b]
Naranjilla	76.40 ± 1.33[c]	21.26 ± 1.35 [c]

The ABTS and DPPH results were presented as the means ± standard deviation. Values in the same column followed by different letters are statistical meaningful at $p < 0.05$ using the ANOVA-one-way analysis followed by the Tukey's Test.

Foods are complex matrices that have many biomolecules and secondary metabolites that may or may not be reactive to the available antioxidant methods. It must be borne in mind that the content of these bioactive compounds in fruits can be modified by the environmental conditions of the crop [56,57]. Therefore, it is important to know the nature of the sample and try to separate the components as much as possible before choosing the method. It would also be appropriate to measure with different methods to look for differences [58,59]. The different bioactive compounds can be separated and identified with different analytical techniques. For example, antioxidant and phenolic compounds can be separated and identified using high performance liquid chromatography (HPLC) coupled with a post column derivatization (PCD). The analyses allow to relate changes in the chemistry of the state of the food directly with the presence or absence of the detected compounds. Another form of analysis of antioxidant and/or polyphenols is to fractionate the sample by semi-preparative HPLC and subsequently test the antioxidant activity of the fractions collected [60]. However, the HPLC-PCD technique is a faster technique and may be more effective for the detection and identification of these bio compounds [61]. Nonetheless, the mathematical model presented in this study for the first time reports that the TAC, TCC and TPC predictions in six fruits of Ecuador, serving as a first screening method. It would be necessary to validate the mathematical model in other fruits to see its effectiveness in order to determine the TAC, TCC, and TPC in other fruits with different qualitative and quantitative phenolic/carotenoid composition. This method can be used as a first screening to determine the TAC, TCC and TPC content but cannot replace an HPLC analysis of the individual components.

4. Conclusions

The mathematical prediction models developed allowed the determination of the TAC, TCC and TPC in araza, Andean blueberry, goldenberry, naranjilla, tamarillo and blackberry in a fast, precise and non-destructive way. These models allow reducing the cost of analysis compared to traditional chemical methods, avoiding the use of solvents and other materials for sample preparation. These satisfactory mathematical models resulted in a prediction of the TAC content of blackberry and Andean blueberry. The TCC was successfully predicted for araza, naranjilla, tamarillo and goldenberry, whereas the TPC was correctly predicted for araza, blackberry, Andean blueberry, tamarillo and goldenberry. The models showed a good statistical calibration as a result of their high determination coefficients R^2, ranging between 0.75 and 0.98. There is a high correlation between experimental and predicted data from the mathematical models. The prediction errors obtained ranged between 0.72% and 6.94% and were kept within the limits according to the Horwitz criterium CV of 5–16%. This method can be used as a routine system susceptible of being evaluated using laboratory systematic methods.

Author Contributions: Conceptualization, W.L., I.S., I.A., B.B., B.O., and W.C.; methodology, W.L., I.S., I.A., B.B., B.O., and W.C.; investigation, W.L., I.S., I.A., B.B., and B.O.; writing—original draft preparation, X.X.; writing—review and editing, W.L.; I.S.; I.A., and W.C.

Funding: The authors gratefully acknowledge that the successful execution of this research work was achieved thanks to the key funding support of the Project 21005270032 belonging to the Strenghthening Institutional Program of the Ecuadorian National Institute of Agricultural Research (INIAP).

Acknowledgments: The kind cooperation of the Laboratory of Food Analysis and Research Service (LSAIA) and the Nutrition and Quality Department of the Santa Catalina Experimentation Station (EESC). W. Carrillo thanks Universidad Técnica de Babahoyo (UTB).

Conflicts of Interest: The authors declare no conflict of interest.

References

1. Moldovan, B.; David, L.; Chişbora, C.; Cimpoiu, C. Degradation kinetics of anthocyanins from European cranberrybush (*Viburnum opulus* L.) fruit extracts. Effects of temperature, pH and storage solvent. *Molecules* **2012**, *17*, 11655–11666. [CrossRef] [PubMed]

2. Kaur, C.; Kapoor, H.C. Antioxidants in fruits and vegetables–The millennium's health. *Int. J. Food Sci. Technol.* **2001**, *36*, 703–725. [CrossRef]

3. Wu, X.; Cheng, J.; Wang, X. Dietary antioxidants: Potential anticancer agents. *Nutr. Cancer* **2017**, *69*, 521–533. [CrossRef] [PubMed]

4. Legua, P.; Forner-Giner, M.Á.; Nuncio-Jáuregui, N.; Hernández, F. Polyphenolic compounds, anthocyanins and antioxidant activity of nineteen pomegranate fruits: A rich source of bioactive compounds. *J. Funct. Foods* **2016**, *23*, 628–636. [CrossRef]

5. Cevallos-Casals, B.A.; Cisneros-Zevallos, L. Stability of anthocyanin-based aqueous extracts of Andean purple corn and red-fleshed sweet potato compared to synthetic and natural colorants. *Food Chem.* **2004**, *86*, 69–77. [CrossRef]

6. Liang, L.; Wu, X.; Zhao, T.; Zhao, J.; Li, F.; Zou, Y.; Yang, L. In vitro bioaccessibility and antioxidant activity of anthocyanins from mulberry (*Morus atropurpurea* Roxb.) following simulated gastro-intestinal digestion. *Food Res. Int.* **2012**, *46*, 76–82. [CrossRef]

7. Osorio, C.; Hurtado, N.; Dawid, C.; Hofmann, T.; Heredia-Mira, F.J.; Morales, A.L. Chemical characterisation of anthocyanins in tamarillo (*Solanum betaceum* Cav.) and Andes berry (*Rubus glaucus* Benth.) fruits. *Food Chem.* **2012**, *132*, 1915–1921. [CrossRef]

8. Benkeblia, N.; Tennant, D.P.F.; Jawandha, S.K.; Gill, P.S. Preharvest and harvest factors influencing the postharvest quality of tropical and subtropical fruits. In *Postharvest Biology and Technology of Tropical and Subtropical Fruits*; Yahia, E., Ed.; Woodhead Publishing: Philadelphia, PA, USA, 2011; pp. 112–142.

9. Khalifa, I.; Zhu, W.; Li, K.K.; Li, C.M. Polyphenols of mulberry fruits as multifaceted compounds: Compositions, metabolism, health benefits, and stability—A structural review. *J. Funct. Foods* **2018**, *40*, 28–43. [CrossRef]

10. Watanabe, T.; Nakamura, N.; Ota, N.; Shiina, T. Estimation of Changes in Mechanical and Color Properties from the Weight Loss Data of "Shine Muscat" Fruit during Storage. *J. Food Quality* **2018**, *6*, 7258029. [CrossRef]

11. Fratianni, A.; Irano, M.; Panfili, G.; Acquistucci, R. Estimation of color of durum wheat. Comparison of WSB, HPLC, and reflectance colorimeter measurements. *J. Agric. Food Chem.* **2005**, *53*, 2373–2378. [CrossRef]

12. Acosta-Montoya, Ó.; Vaillant, F.; Cozzano, S.; Mertz, C.; Pérez, A.M.; Castro, M.V. Phenolic content and antioxidant capacity of tropical highland blackberry (*Rubus adenotrichus* Schltdl.) during three edible maturity stages. *Food Chem.* **2010**, *119*, 1497–1501. [CrossRef]

13. Almeida, J.; Asís, R.; Molineri, V.N.; Sestari, I.; Lira, B.S.; Carrari, F.; Rossi, M. Fruits from ripening impaired, chlorophyll degraded and jasmonate insensitive tomato mutants have altered tocopherol content and composition. *Phytochemistry* **2015**, *111*, 72–83. [CrossRef] [PubMed]

14. Sdiri, S.; Navarro, P.; Monterde, A.; Benabda, J.; Salvador, A. Effect of postharvest degreening followed by a cold-quarantine treatment on vitamin C, phenolic compounds and antioxidant activity of early-season citrus fruit. *Postharvest Biol. Technol.* **2012**, *65*, 13–21. [CrossRef]

15. Brannan, R.G.; Wang, G. Effect of frozen storage on polyphenol oxidase, antioxidant content, and color of pawpaw (*Asimina triloba* [L.] Dunal) fruit pulp. *J. Food Res.* **2017**, *6*, 93–101. [CrossRef]

16. Abdullah, M.Z.; Guan, L.C.; Lim, K.C.; Karim, A.A. The applications of computer vision system and tomographic radar imaging for assessing physical properties of food. *J. Food Eng.* **2004**, *61*, 125–135. [CrossRef]

17. Hue, C.; Brat, P.; Gunata, Z.; Samaniego, I.; Servent, A.; Morel, G.; Davrieux, F. Near infra-red characterization of changes in flavan-3-ol derivatives in cocoa (*Theobroma cacao* L.) as a function of fermentation temperature. *J. Agric. Food Chem.* **2014**, *62*, 10136–10142. [CrossRef] [PubMed]

18. Leon, K.; Mery, D.; Pedreschi, F.; Leon, J. Color measurement in L∗ a∗ b∗ units from RGB digital images. *Food Res. Int.* **2006**, *39*, 1084–1091. [CrossRef]

19. Ferrer, A.; Remón, S.; Negueruela, A.I.; Oria, R. Changes during the ripening of the very late season Spanish peach cultivar Calanda: Feasibility of using CIELAB coordinates as maturity indices. *Sci. Hortic.* **2005**, *105*, 435–446. [CrossRef]

20. Hurtado, N.H.; Morales, A.L.; González-Miret, M.L.; Escudero-Gilete, M.L.; Heredia, F.J. Colour, pH stability and antioxidant activity of anthocyanin rutinosides isolated from tamarillo fruit (*Solanum betaceum* Cav.). *Food Chem.* **2009**, *117*, 88–93. [CrossRef]

21. Abdullah, M.Z.; Guan, L.C.; Azemi, B.M. Stepwise discriminant analysis for colour grading of oil palm using machine vision system. *Food Bioprod. Process.* **2001**, *79*, 223–231. [CrossRef]

22. Cefola, M.; D'Antuono, I.; Pace, B.; Calabrese, N.; Carito, A.; Linsalata, V.; Cardinali, A. Biochemical relationships and browning index for assessing the storage suitability of artichoke genotypes. *Food Res. Int.* **2012**, *48*, 397–403. [CrossRef]

23. Lee, B.; Seo, J.D.; Rhee, J.K.; Kim, C.Y. Heated apple juice supplemented with onion has greatly improved nutritional quality and browning index. *Food Chem.* **2016**, *201*, 315–319. [CrossRef] [PubMed]

24. Jha, S.N.; Narsaiah, K.; Basediya, A.L.; Sharma, R.; Jaiswal, P.; Kumar, R.; Bhardwaj, R. Measurement techniques and application of electrical properties for nondestructive quality evaluation of foods—A review. *J. Food Sci. Technol.* **2011**, *48*, 387–411. [CrossRef] [PubMed]

25. Kowalski, B.R.; Seasholtz, M.B. Recent developments in multivariate calibration. *J. Chemom.* **1991**, *5*, 129–145. [CrossRef]

26. Memmah, M.M.; Lescourret, F.; Yao, X.; Lavigne, C. Metaheuristics for agricultural land use optimization. A review. *Agron. Sustain. Dev.* **2015**, *35*, 975–998. [CrossRef]

27. Garzón, G.A.; Riedl, K.M.; Schwartz, S.J. Determination of anthocyanins, total phenolic content, and antioxidant activity in Andes berry (*Rubus glaucus* Benth). *J. Food Sci.* **2009**, *74*, C227–C232. [CrossRef] [PubMed]

28. Mejía, D.; Maria, C.; Gaviria, A.; Duque, C.; Lucia, A.; Rengifo, R.; Hernan, A. Changes in physicochemical properties of the fruit of lulo (*Solanum quitoense* Lam.) harvested at three degrees of maturity. *Vitae* **2012**, *19*, 157–165.

29. Gordon, A.; Rodrigues, R.B.; Marx, F.; Papagiannopoulos, M. Antioxidant capacity of tamarillo fruit (*Cyphomandra betacea*). Utilisation of diversity in land use systems: Sustainable and organic approaches to meet human needs. In Proceedings of the Conference on International Agricultural Research for Development, Witzenhausen, Germany, 9–11 October 2007.

30. Ramadan, M.F. Bioactive phytochemicals, nutritional value, and functional properties of cape gooseberry (*Physalis peruviana*): An overview. *Food Res. Int.* **2011**, *44*, 1830–1836. [CrossRef]

31. Hernández, M.S.; Martínez, O.; Fernández-Trujillo, J.P. Behavior of arazá (*Eugenia stipitata* Mc Vaugh) fruit quality traits during growth, development and ripening. *Sci. Hortic.* **2007**, *111*, 220–227. [CrossRef]

32. Ribera, A.E.; Reyes-Diaz, M.; Alberdi, M.; Zuñiga, G.E.; Mora, M.L. Antioxidant compounds in skin and pulp of fruits change among genotypes and maturity stages in highbush blueberry (*Vaccinium corymbosum* L.) grown in southern Chile. *J. Soil Sci. Plant Nutr.* **2010**, *10*, 509–536. [CrossRef]

33. Acosta-Quezada, P.G.; Raigón, M.D.; Riofrío-Cuenca, T.; García-Martínez, M.D.; Plazas, M.; Burneo, J.I.; Prohens, J. Diversity for chemical composition in a collection of different varietal types of tree tomato (*Solanum betaceum* Cav.), an Andean exotic fruit. *Food Chem.* **2015**, *169*, 327–335. [CrossRef] [PubMed]

34. Acosta, Ó.; Pérez, A.M.; Vaillant, F. Chemical characterization, antioxidant properties, and volatile constituents of naranjilla (*Solanum quitoense* Lam.) cultivated in Costa Rica. *Archiv. Latinoam. Nutr.* **2009**, *59*, 88.

35. Fischer, G.; Herrera, A.; Almanza, P.J. Cape gooseberry (*Physalis peruviana* L.). In *Stharvest Biology and Technology of Tropical and Subtropical Fruits. Acai to Citrus*; Yahia, E.M., Ed.; Woodhead Publishing: Cambridge, UK, 2011; pp. 374–396.

36. Tehranifar, A.; Zarei, M.; Nemati, Z.; Esfandiyari, B.; Vazifeshenas, M.R. Investigation of physicochemical properties and antioxidant activity of twenty Iranian pomegranate (*Punica granatum* L.) cultivars. *Sci. Hortic.* **2010**, *126*, 180–185. [CrossRef]

37. Rapisarda, P.; Fanella, F.; Maccarone, E. Reliability of analytical methods for determining anthocyanins in blood orange juices. *J. Agric. Food Chem.* **2000**, *48*, 2249–2252. [CrossRef] [PubMed]

38. Leong, S.Y.; Oey, I. Effects of processing on anthocyanins, carotenoids and vitamin C in summer fruits and vegetables. *Food Chem.* **2012**, *133*, 1577–1587. [CrossRef]

39. Georgé, S.; Brat, P.; Alter, P.; Amiot, M.J. Rapid determination of polyphenols and vitamin C in plant-derived products. *J. Agric. Food Chem.* **2005**, *53*, 1370–1373. [CrossRef] [PubMed]

40. Piñuel, L.; Vilcacundo, E.; Boeri, P.; Barrio, D.A.; Morales, D.; Pinto, A.; Morán, R.; Samaniego, I.; Carrillo, W. Extraction of protein concentrate from red bean (*Phaseolus vulgaris* L.): Antioxidant activity and inhibition of lipid peroxidation. *J. Appl. Pharm. Sci.* **2019**, *9*, 1–14.

41. Cuellar, F.A.; Ariza, E.; Anzola, C.; Restrepo, P. Research of antioxidant capacity of araza (*Eugenia stipitata* Mc Vaugh) during the ripening. *Rev. Colomb. Quim.* **2013**, *42*, 21–28.

42. Carvalho, C.P.; Betancur, J.A. Quality characterization of Andean blackberry fruits (*Rubus glaucus* Benth.) in different maturity stages in Antioquia, Colombia. *Agron. Colomb.* **2015**, *33*, 74–83. [CrossRef]

43. Peña, A.C.; González, M.L.; Hernández, M.S.; Novoa, C.; Quicazán, M.C.; Fernández-Trujillo, J.P. Evaluation of arazá (*Eugenia stipitate* Mc Vaugh) fruit after wet behavior postharvest conditioning operations. *Rev. Colomb. Cien. Hortic.* **2007**, *1*, 182–188. [CrossRef]

44. Kalt, W.; McRae, K.B.; Hamilton, L.C. Relationship between surface color and other maturity indices in wild lowbush blueberries. *Can. J. Plant Sci.* **1995**, *75*, 485–490. [CrossRef]

45. Mertz, C.; Gancel, A.L.; Gunata, Z.; Alter, P.; Dhuique-Mayer, C.; Vaillant, F.; Brat, P. Phenolic compounds, carotenoids and antioxidant activity of three tropical fruits. *J. Food Comp. Anal.* **2009**, *22*, 381–387. [CrossRef]

46. Horwitz, W.; Kamps, L.R.; Boyer, K.W. Quality assurance in the analysis of foods and trace constituents. *J. Assoc. Off. Anal. Chem.* **1980**, *63*, 1344–1354. [PubMed]

47. Vasco, C.; Riihinen, K.; Ruales, J.; Kamal-Eldin, A. Chemical composition and phenolic compound profile of mortiño (*Vaccinium floribundum* Kunth). *J. Agric. Food Chem.* **2009**, *57*, 8274–8281. [CrossRef] [PubMed]

48. Contreras-Calderón, J.; Calderón-Jaimes, L.; Guerra-Hernández, E.; García-Villanova, B. Antioxidant capacity, phenolic content and vitamin C in pulp, peel and seed from 24 exotic fruits from Colombia. *Food Res. Int.* **2011**, *44*, 2047–2053. [CrossRef]

49. Vasco, C.; Ruales, J.; Kamal-Eldin, A. Total phenolic compounds and antioxidant capacities of major fruits from Ecuador. *Food Chem.* **2008**, *111*, 816–823. [CrossRef]

50. Laverde-Acurio, L.J. Estudio de las condiciones óptimas para la obtención de jugo clarificado de arazá (*Eugenia stipitata*), mediante procesos enzimático y membranario. Bachelor's Thesis, Escuela Politécnica Nacional, Quito, Ecuador, 2010. Available online: https://bibdigital.epn.edu.ec/handle/15000/2558 (accessed on 6 November 2010).

51. Torres, N. Determinación del potencial nutritivo y nutracéutico de cuatro cultivares de tomate de árbol (*Solanum betaceum* Cav.). Bachelor's Thesis, Escuela Superior Politécnica de Chimborazo, Riobamba, Ecuador, 2006. Available online: http://repositorio.iniap.gob.ec/handle/41000/921 (accessed on 1 June 2015).
52. Cerón, I.; Higuita, J.; Cardona, C. Antioxidant capacity and total phenolic content of three fruits from Andean region. *Vector* **2010**, *5*, 17–26.
53. Tupuna, D. Obtención de jugo concentrado clarificado de mortiño mediante el uso de tecnología de membranas. Bachelor's Thesis, Facultad de Ingeniería Agroindustrial, Escuela Politécnica Nacional, Quito, Ecuador, 2012. Available online: https://bibdigital.epn.edu.ec/handle/15000/4947 (accessed on 12 September 2012).
54. Espin, S.; González-Manzano, S.; Taco, V.; Poveda, C.; Ayuda-Durán, B.; González-Paramas, A.M.; Santos-Buelga, C. Phenolic composition and antioxidant capacity of yellow and purple-red Ecuadorian cultivars of tree tomato (*Solanum betaceum* Cav.). *Food Chem.* **2016**, *194*, 1073–1080. [CrossRef]
55. Montalvo, D.A. Evaluación de la calidad postcosecha de las accesiones seleccionadas de mora de castilla (*Rubus glaucus* Benth) provenientes de las provincias de Tungurahua y Bolívar. Bachelor's Thesis, Facultad de Ingeniería Química y Agroindustria, Escuela Politécnica Nacional, Quito, Ecuador, 2010. Available online: https://bibdigital.epn.edu.ec/bitstream/15000/2653/1/CD-3336.pdf (accessed on 4 January 2011).
56. Do Socorro, M.R.; Alves, R.E.; de Brito, E.S.; Pérez-Jiménez, J.; Saura-Calixto, F.; Mancini-Filho, J. Bioactive compounds and antioxidant capacities of 18 non-traditional tropical fruits from Brazil. *Food Chem.* **2010**, *121*, 996–1002.
57. Pisoschi, A.M.; Pop, A.; Cimpeanu, C.; Predoi, G. Antioxidant capacity determination in plants and plant-derived products: A review. *Oxid. Med. Cell. Longev.* **2016**. [CrossRef]
58. Banerjee, J.; Singh, R.; Vijayaraghavan, R.; MacFarlane, D.; Patti, A.F.; Arora, A. Bioactive from fruit processing wastes: Green approaches to valuable chemicals. *Food Chem.* **2017**, *225*, 10–22. [CrossRef] [PubMed]
59. Alam, M.N.; Bristi, N.J.; Rafiquzzaman, M. Review on in vivo and in vitro methods evaluation of antioxidant activity. *Saudi Pharm J.* **2013**, *21*, 143–152. [CrossRef] [PubMed]
60. Cádiz-Gurrea, M.; Lozano-Sanchez, J.; Contreras-Gámez, M.; Legeai-Mallet, L.; Fernández-Arroyo, S.; Segura-Carretero, A. Isolation, comprehensive characterization and antioxidant activities of Theobroma cacao extract. *J. Funct. Foods* **2014**, *10*, 485–498. [CrossRef]
61. Acquaviva, A.; Jones, A.; Dennis, G.R.; Shalliker, R.A.; Soliven, A. Phenolic profiling of complex tea samples via simultaneous multiplexed detection employing reaction flow HPLC columns and colorimetric post column derivatisation. *Microchem. J.* **2018**, *138*, 533–539. [CrossRef]

Article

Bioprocessed Production of Resveratrol-Enriched Rice Wine: Simultaneous Rice Wine Fermentation, Extraction, and Transformation of Piceid to Resveratrol from *Polygonum cuspidatum* Roots

Kai-Ruei Yang [1,†], Hui-Chuan Yu [2,†], Chun-Yung Huang [1], Jen-Min Kuo [1], Cheng Chang [2], Chwen-Jen Shieh [2,*] and Chia-Hung Kuo [1,*]

[1] Department of Seafood Science, National Kaohsiung University of Science and Technology, Kaohsiung 811, Taiwan
[2] Biotechnology Center, National Chung Hsing University, Taichung 402, Taiwan
*   Correspondence: cjshieh@nchu.edu.tw (C.-J.S.); kuoch@nkust.edu.tw (C.-H.K.);
    Tel.: +886-4-2284-0450 (ext. 5121) (C.-J.S.); -886-7-361-7141 (ext. 23646) (C.-H.K.)
†   These authors contributed equally to this work.

Received: 24 June 2019; Accepted: 10 July 2019; Published: 15 July 2019

Abstract: A new bioprocess to produce resveratrol-enriched rice wine was established and the effects of adding *Polygonum cuspidatum* root powder to rice wine fermentation were investigated. In this new process, piceid and resveratrol were extracted from *P. cuspidatum* roots to rice wine and piceid was converted to resveratrol by β-glucosidase during fermentation. After 10 days co-fermentation, rice wine with high levels of resveratrol was obtained, which contained ~14% (*v/v*) ethanol, 122 mg/L piceid, and 86 mg/L resveratrol. The resveratrol-enriched rice wine had enhanced antioxidant activity with significantly stronger 2,2-diphenyl-1-picrylhydrazyl (DPPH) radical scavenging activity, ferric ion reducing power, and ferrous ion chelating capability. Ultrafiltration (UF) was employed in this study using hollow fibers to clarify the end product, increase shelf life without heat treatment, and maintain the quality of the phenolic compounds. The boiled and UF-treated rice wine were evaluated for ethanol, piceid, resveratrol, clarity, aerobic plate count, total acidity, pH, reducing sugars, and amino acids. The quality of the resveratrol-enriched rice wine was maintained after four weeks storage at normal refrigeration temperatures.

Keywords: rice wine; piceid; resveratrol; *Polygonum cuspidatum*; antioxidant activity; ultrafiltration; clarification

1. Introduction

Resveratrol (3,5,4-trihydroxystilbene) is a polyphenol that exists in several plants, including bilberry, blueberry, cranberry, grape, and peanut [1–3]. Resveratrol possesses physiological functions reported to inhibit the migration and metastasis of cancer cells [4], as well as antioxidative, anti-melanoma [5], fatty-liver-reducing [6], anti-obesity [7], anti-inflammatory, cardioprotective, neuroprotective, and antitumor properties [8]. Resveratrol from grape juice [9], grape seeds [10], grape skins [11], and red wine [12] has been extensively studied. It is generally agreed that moderate consumption of red wine can prevent cardiovascular disease. The resveratrol in red wine is believed to be a key molecule since it inhibits low-density lipoprotein oxidation and platelet aggregation in blood [13,14]. Resveratrol is mainly present in grape skins and its concentration in grape pulp is low or absent altogether. Red wine is made from pomace and fermentation occurs together with grape skins, which gives the wine higher amounts of resveratrol than white wine [15,16]. Despite this, resveratrol concentration in red wine is relatively low, ranging from around 1.6 to 3.6 mg/L [17].

Polygonum cuspidatum is a perennial plant belonging to the Polygonaceae family. Dried *P. cuspidatum* roots, also called Hu Zhang in Chinese, are herbal medicine for the treatment of cough, arthralgia, chronic bronchitis, jaundice, amenorrhea, hypertension, and hypercholesterolemia [18]. The main phenolic compounds extracted from *P. cuspidatum* roots have been identified as resveratrol, piceid, and emodin [19]. Emodin has been shown to possess anti-inflammatory, antibacterial, antiglycation, and antineoplastic activities [20–23]. However, emodin is more hydrophobic and only small amounts of emodin can be extracted at low ethanol concentrations compared to piceid and resveratrol [24]. In plants, resveratrol is usually present in piceid (resveratrol 3-β-mono-D-glucoside), its glycoside derivative. Extracts from *P. cuspidatum* roots contain high amounts of piceid, which can be converted to resveratrol when treated with β-glucosidase or cellulase or fermented by *Aspergillus oryzae* [25–27]. Rice wine is an alcoholic beverage made from rice, traditionally consumed in East Asia, Southeast Asia, and South Asia [28]. Qu is usually used as a starter in rice wine fermentation [29]. Qu is rich in a wide variety of microorganisms, such as the filamentous molds *A. oryzae* and *Rhizopus oryzae* and the amylolytic yeast *Saccharomycopsis fibuligera*, as well as various enzymes, including amylase, glucoamylase, protease, and phosphatase [30,31]. *A. oryzae* and *R. oryzae* are responsible for producing amylase during fermentation; these two microorganisms can also produce β-glucosidase [32,33].

Functional beer, a new product recently obtained via co-fermentation with medicinal herbs, has been commercialized in Japan [34], which has prompted interest in developing a high value rice wine with health benefits. Most medicinal herbs are rich in antioxidants, especially phenolic compounds. Therefore, wines made from medicinal herbs are rich in natural antioxidants to enhance the health functions of fermented wines. In recent years, resveratrol has become a candidate nutritional substance for cancer prevention widely available as a botanical dietary supplement. To date, *P. cuspidatum* roots are used as a commercial source of resveratrol. However, the conversion of excessive piceid in *P. cuspidatum* to resveratrol is still necessary. As such, the addition of *P. cuspidatum* roots during rice wine fermentation to produce resveratrol-enriched rice wine was investigated. In this new process, as depicted in Figure 1, resveratrol can be extracted and piceid converted to resveratrol by β-glucosidase during fermentation.

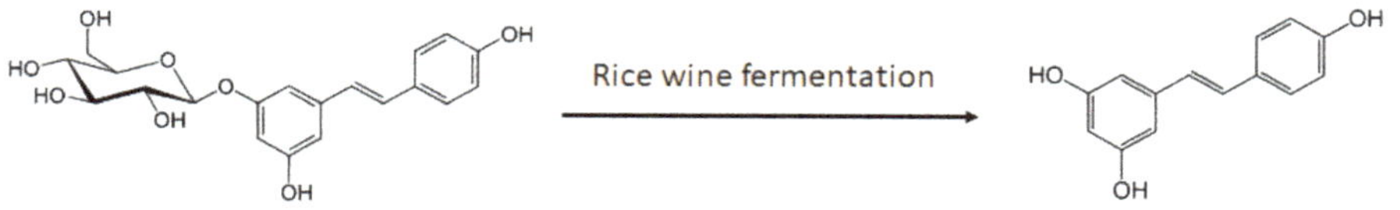

Figure 1. Scheme representing transformation of piceid to resveratrol during rice wine fermentation.

Sterilization by boiling is commonly used to kill microorganisms and deactivate the enzymes in rice wine to protect it against rancidity and deterioration. However, boiling at high temperatures might increase turbidity, cause heavy browning, or decompose bioactive compounds. Ultrafiltration (UF) is a new technology widely used by the food industry in recent years to remove enzymes, microorganisms, and turbidity contents. The separation principle of the UF membrane is based on mechanical filtration, driven by the pressure difference between the internal and external sides of the membrane [35–38]. Compared to boiling, the advantages of UF are no phase changes, operation at low temperatures, low energy consumption, simple operation, and better preservation of the original flavor and nutrients. At present, many studies use ultrafiltration for juice clarification [39,40], but the application of UF in wine sterilization is still limited.

The objective of this study was to develop a new brewing process for resveratrol-enriched rice wine. The effects of *P. cuspidatum* on the composition characteristics of rice wine were investigated. The antioxidant activity of rice wine was determined based on the scavenging activity of 2,2-diphenyl-1-picrylhydrazyl (DPPH) radicals, ferrous ion chelating activity, and reducing activity. Finally, ultrafiltration was employed to remove enzymes, microorganisms, and turbidity contents from the resveratrol-enriched rice wine.

2. Materials and Methods

2.1. Materials

Polished rice (11.9% moisture, 82.4% carbohydrate, 5.7% protein, and 0% lipid) from the 2017 crop was purchased from the Hua-Tung Rice Co., Ltd. (Hualien, Taiwan). Dried *P. cuspidatum* roots were purchased from local Chinese herbal medicine stores and ground into powder with ~0.62 mm particles. Qu was purchased from Yong Xin Jiuqu Co., Ltd. (Changhua, Taiwan). Piceid and *p*-nitrophenyl-β-D-glucopyranoside (PNG) was purchased from Sigma-Aldrich (MO, USA). Resveratrol was purchased from Changsha Nutramax Biotechnology (Changsha, China). DPPH (2,2-Diphenyl-1-picrylhydrazyl) and *o*-phthaldialdehyde were purchased from Alfa Aesar (Tewksbury, MA, USA). Finally, 3,5-dinitrosalicylic acid and ferrozine were purchased from Acros (Morris Plains, NJ, USA). Unless otherwise noted, all reagents and chemicals were of analytical grade.

2.2. Rice Wine Production

Rice and water at a ratio of 1:1 were steam-cooked for 40 min. The moisture content of the cooked rice was 45.42%. After cooling to 30 °C, Qu was added to the rice at a ratio of 1:200 (Qu: steam-cooked rice; *w/w*) and mixed well. Then, 100 g of the mixture was added to a 500 mL screw capped glass jar and the mixture was cultured at 30 °C for 2 days. Varying amounts of *P. cuspidatum* root powder (1%, 3%, or 5% based on the weight of the steam-cooked rice) and 100 mL sterile water were then added to the jar for simultaneous extraction and transformation of piceid to resveratrol from *P. cuspidatum* root powder during rice wine fermentation. Ethanol, piceid, and resveratrol concentrations were measured periodically during fermentation.

2.3. Clarification and Sterilization by Ultrafiltration

Rice wine fermented with 5% *P. cuspidatum* for 10 days was used for UF processing. The rice wine was centrifuged at 10,000 rpm for 5 min and the supernatant filtered through Whatman No. 1 filter paper. The rice wine was then ultrafiltered by a tangential flow filtration system (MAP-TFF, Lefo Science, Taipei, Taiwan) equipped with a hollow-fiber filter module with a molecular weight cut-off of 3 kDa or 10 kDa (MicroKros, Spectrum Labs, Rancho Dominguez, CA, USA). The rice wine was stored at 4 °C to test its shelf life. Samples were analyzed for ethanol, piceid, and resveratrol concentrations, as well as total acidity, turbidity, reducing sugars, amino acids, and aerobic plate count (APC). The control experiment, sterilization by boiling, was conducted by placing the centrifuged supernatant in a boiling water bath for 20 min.

2.4. Antioxidative Properties of Resveratrol-Enriched Rice Wine

The DPPH radical scavenging activity was measured according to the method of Huang et al. [41], with some minor modifications. Briefly, 0.5 mL of the sample was added to 0.5 mL 0.1 mM freshly prepared DPPH solution (in ethanol). After DPPH radicals transferred hydrogen to antioxidative agents, the solution lightened in color at 517 nm due to the reduction in optical absorbance. The mixture was shaken vigorously for 1 min then left to stand for 30 min in the dark at room temperature. The absorbance of all sample solutions was measured at 517 nm using a UV/VIS spectrophotometer (Hitachi U-2900, Tokyo, Japan). The DPPH radical scavenging activity was calculated using the following equation:

$$\text{Scavenging activity (\%)} = (1 - (A_{sample})/A_{control}) \times 100 \tag{1}$$

The reducing power assay was measured according to the method of Conde et al. [42], with some minor modifications. Briefly, 1 mL of sample was added to 1 mL of phosphate buffer (200 mM, pH 6.6) and 1 mL of 1% potassium ferricyanide. The solution was allowed to react for 20 min at 50 °C, and then 1 mL of 10% trichloroacetic acid (in ethanol) was added. The reactant was centrifuged at 10,000 rpm for 10 min. Then 0.5 mL of the upper layer of the solution was mixed with 0.5 mL of distilled water and 0.1 mL of 0.1% ferric chloride in test tubes. After 10 min of reaction, the resulting solution was measured at 700 nm. Increased absorbance (A700) of the reaction mixture indicated increased reducing power. The standard curve was linear between 20 and 100 ppm vitamin C. Results are expressed in ppm vitamin C. Additional dilution was needed if the measured absorbance value was over the linear range of the standard curve.

The ferrous ion chelating power was measured according to the method of Wu et al. [43], with some minor modifications. Briefly, 1 mL of sample was added to 0.1 mL of 2 mM ferrous chloride and 3.7 mL of methanol. The reaction was started by the addition of 5 mM ferrozine (0.2 mL), and the mixture was shaken vigorously before being left to stand at room temperature for 10 min. Absorbance of the resulting solution was measured at 562 nm. The chelating power was calculated according to Equation (1).

2.5. Analysis

The ethanol concentration was quantified using a Thermo Quest Trace 2000 gas chromatograph equipped with a flame ionization detector and an MXT-WAX capillary column (30 m × 0.28 mm i.d.; film thickness 0.25 μm; RESTEK, Bellefonte, PA, USA). Injector and detector temperatures were set at 200 °C and 250 °C, respectively. Initial temperature of the column oven was set at 40 °C for 2 min, and then increased to 215 °C at 20 °C min^{-1}. Pure helium was used as a carrier gas at a flow rate of 0.5 mL min^{-1}. Piceid and resveratrol were assayed by injecting 20 μL of the sample into an HPLC system, consisting of a Hitachi L-2130 HPLC pump and a Hitachi L-2420 UV/VIS detector (Hitachi, Tokyo, Japan), using an Inertsil ODS-3 column (5 μM, 250 mm × 4.6 mm). Deionized water and methanol containing 0.1% acetic acid were used for gradient elution from 10 to 100% methanol for 20 min, followed by elution at 100% methanol for 5 min. The flow rate was set at 1.0 mL min^{-1}. The UV detector was set at a wavelength of 303 nm. The standard curves for piceid and resveratrol were linear over the range 25–125 μg/mL (R^2 = 0.9987) and 16–80 μg/mL (R^2 = 0.9995), respectively. Piceid and resveratrol levels in the samples were quantified by comparing their retention times and peak areas with those of the standards. Samples were diluted if the concentrations fell outside the standard curve ranges. β-Glucosidase was determined using PNG as the substrate. Briefly, 50 μL of sample was added to 50 μL of 20 mM PNG (in 50 mM, pH 4.5 citric buffer) and was incubated for 30 min at 37 °C. The reaction was stopped by the addition of 0.9 mL of 0.1 M sodium carbonate. The release of *p*-nitrophenol, resulting from the β-glucosidase-catalyzed hydrolysis of PNG, was measured by reading the absorbance at 405 nm ($\varepsilon405$ = 17.0 mM^{-1} cm^{-1}). One unit of β-glucosidase activity was defined as the amount of enzyme that liberated 1 μmoL of *p*-nitrophenol per minute. The reducing sugar content of the rice wine was determined by 3,5-dinitrosalicylic acid colorimetry using glucose as the standard [44]. Total acidity of the rice wine was rapidly determined by titrimetry. The amino acid content was determined by *o*-phthaldialdehyde colorimetry using glutamine as the standard [45]. Clarity was determined by measuring the transmittance (T%) at 680 nm using a UV/VIS spectrophotometer (Hitachi U-2900, Tokyo, Japan); distilled water was used as the control.

3. Results and Discussion

3.1. Rice Wine Fermented with P. cuspidatum

Steam-cooked rice mixed with Qu was incubated for 2 days in order to induce the enzymes involved in ethanol fermentation, including amylase, glucoamylase, and protease. After incubation, varying amounts of *P. cuspidatum* root powder (1%, 3%, or 5%) were added with sterile water to the mixture for rice wine fermentation and simultaneous extraction of piceid and resveratrol. The ethanol yield during the period of fermentation is shown in Figure 2. The addition of *P. cuspidatum* decreased the initial ethanol production rate for the first 6 days, compared to the control; however, the ethanol yields were the same after 8 d of fermentation. The final ethanol yields were 14.2%, 14.4%, 14.5%, and 14.1%, for rice wine with 1%, 3%, and 5% *P. cuspidatum* and the control, respectively. The results showed that *P. cuspidatum* did not influence ethanol production.

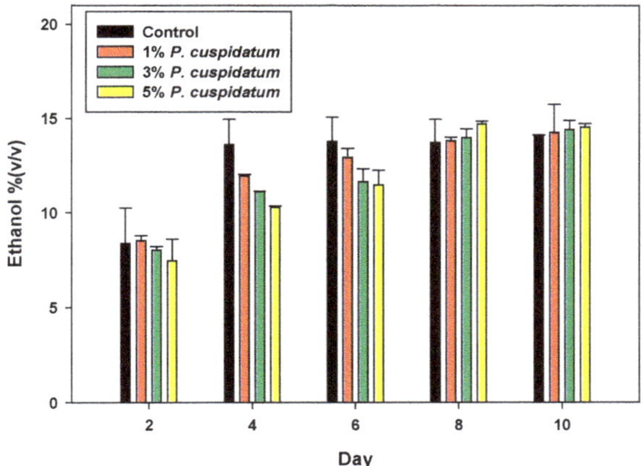

Figure 2. Effect of *P. cuspidatum* roots on ethanol yield during rice wine fermentation. Control was performed without the addition of *P. cuspidatum*.

3.2. Biotransformation of Piceid to Resveratrol during Rice Wine Fermentation

Piceid and resveratrol, which originated from the *P. cuspidatum* roots, were infused into the rice wine during fermentation. The change in piceid yield during fermentation is shown in Figure 3a. The highest piceid concentrations were found in the first 2 days at 73, 161, and 231 mg/L for wine with 1%, 3%, and 5% *P. cuspidatum*, respectively, but after 10 days, these levels decreased to 22, 80, and 122 mg/L. Resveratrol showed a different trend; the change in its yield during fermentation is shown in Figure 3b. Resveratrol concentration was lowest the first 2 days at 10, 22, and 34 mg/L for wine with 1%, 3%, and 5% *P. cuspidatum*, respectively. Unlike piceid, however, resveratrol concentration increased with fermentation time, with the yield increasing gradually for the first 8 days and then levelling off after 10 days. Wang et al. studied the biotransformation of piceid to resveratrol by *A. oryzae* and reported that the yield of trans-resveratrol first increased, and then decreased after it reached its highest value during fermentation [46]. It has been reported that resveratrol increases the life span of yeast in winemaking [47]. The levelling off in resveratrol concentration after 10 days may be due to the converted resveratrol being consumed by yeast. The highest resveratrol yield was 26, 57, and 86 mg/L after 10 days for wine with 1%, 3%, and 5% *P. cuspidatum*, respectively. These results indicate that piceid was converted to resveratrol during rice wine fermentation.

β-Glucosidase catalyzes the hydrolysis of β-O-glucosidic linkages between β-D-glucose and an aglycone [48,49]. Therefore, β-glucosidase activity was examined during rice wine fermentation. As shown in Figure 4, β-glucosidase activity increased with the amount of *P. cuspidatum* supplementation. After 10 days fermentation, the β-glucosidase activity was 5.3, 7.6, and 10.2 U/L for wine with 1%, 3%, and 5% *P. cuspidatum*, respectively. The results indicate that β-glucosidase activity was induced by *P. cuspidatum* during fermentation. The mole conversion of piceid to resveratrol was defined as the increase of resveratrol mole per the decrease of piceid mole from days 2 to 10. The mole conversions for wine with 1%, 3%, and 5% *P. cuspidatum* were 56%, 73%, and 82%, respectively. These results indicate that the mole conversion is directly related to β-glucosidase activity. β-Glucosidase from *A. niger* has been used to hydrolyze piceid to produce resveratrol in red wine, which increases resveratrol to 75% [50], but the overall content of resveratrol is still low (~2.8 mg/L).

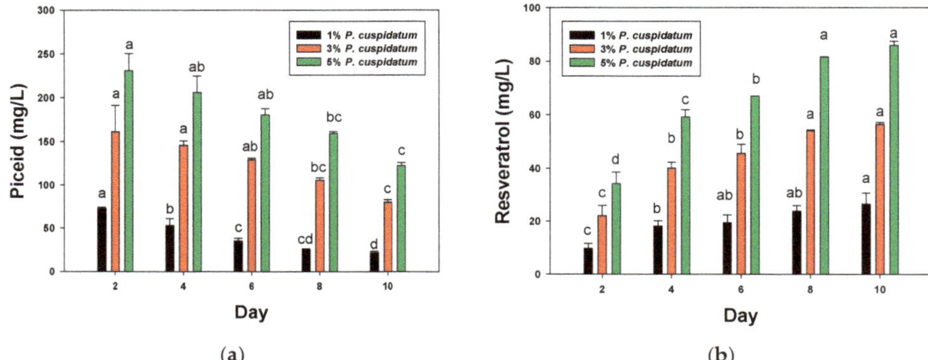

Figure 3. Effect of *P. cuspidatum* roots on the (**a**) piceid and (**b**) resveratrol yields during rice wine fermentation. [a,b,c,d] When the bars of the same color have significant differences at *p*-value less than 0.05, homogeneous groups in each variable are identified by the same superscript letter according to the LSD (least significant difference) test.

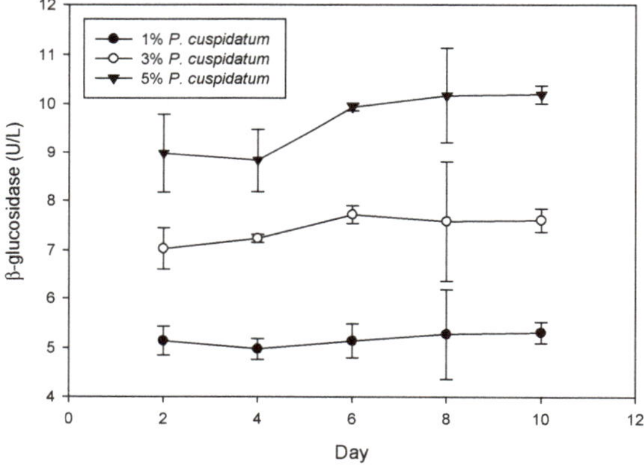

Figure 4. Effect of *P. cuspidatum* roots on the β-glucosidase activity during rice wine fermentation.

3.3. Antioxidant Capacity of Resveratrol-Enriched Rice Wine

Oxygen is the basic component needed to maintain most life, but oxygen can form reactive oxygen species (ROS) or free radicals, such as superoxide anion radicals, hydroxyl radicals, hydrogen peroxide, singlet oxygen, or nitric oxide [51]. Excessive free radicals or ROS are harmful to the human body because these components are highly chemically reactive and may damage cellular DNA, proteins, nucleic acids, and lipids, thereby destroying the normal functions of cells, tissues, and organs. The importance of anti-oxidative components is due to their ability to reduce the damage to free radical-mediated cells, tissues, and organs within the human body. Polyphenols possess a well-known antioxidant capacity because of their ability to scavenge free radicals, inhibit free radical formation, and chelate with metals to protect cells from oxidative damage and further prevent the occurrence of cancer and cardiovascular disease [52]. Therefore, the aim of this section was to demonstrate the antioxidant capacity of resveratrol-enriched rice wine.

The DPPH radical scavenging capacities of rice wine, and rice wine fermented with 1%, 3%, and 5% *P. cuspidatum* root were 5.12%, 29.70%, 30.42%, and 37.18%, respectively (Table 1). The DPPH radical scavenging capacity of the rice wine fermented with *P. cuspidatum* was higher than the control. DPPH radical scavenging activity increased with the amount of *P. cuspidatum* supplementation.

The higher percentage values of DPPH radical scavenging capacity in rice wine fermented with *P. cuspidatum* indicate that resveratrol-enriched rice wine scavenges free radicals in vitro, consistent with the study results for total amounts of resveratrol. Similarly, the DPPH radical scavenging capacity for 250 mg/L of *P. cuspidatum* extract was 22.22% [5], which is lower than the resveratrol-enriched rice wine. This might be due to the conversion of piceid to resveratrol during rice wine fermentation.

Table 1. DPPH free radical scavenging activity, ferrous ion chelating activity and reducing power of resveratrol-enriched rice wine. Control was performed without the addition of *P. cuspidatum*.

Group	DPPH Radical Scavenging Activity (%)	Reducing Power (Vitamin C Equivalent ppm)	Ferrous-Ion Chelating Activity (%)
Control	5.12 ± 2.70 [c]	6.93 ± 0.20 [d]	69.36 ± 0.83 [c]
1% *P. cuspidatum*	29.70 ± 0.11 [b]	107.29 ± 6.43 [c]	73.05 ± 1.36 [b]
3% *P. cuspidatum*	30.42 ± 1.60 [b]	189.63 ± 6.94 [b]	79.14 ± 0.61 [a]
5% *P. cuspidatum*	37.18 ± 6.05 [a]	293.92 ± 3.47 [a]	80.91 ± 1.28 [a]

[a,b,c,d] When there are significant differences at *p*-value less than 0.05, homogeneous groups in each variable are identified by the same superscript letter according to the LSD test. DPPH: 2,2-Diphenyl-1-picrylhydrazyl.

Reducing power is another important indicator of antioxidant activity in natural components. The reducing capacity of a compound may serve as a significant indicator of its potential antioxidant activity and indicates that these compounds are electron donors [53]. The quantification of reducing power is based on the antioxidant's reaction with potassium ferricyanide (Fe^{3+}) to form potassium ferrocyanide (Fe^{2+}). This then reacts with ferric chloride to form a ferric ferrous complex that has a maximum absorption at 700 nm. As shown in Table 1, the reducing powers of rice wine, and rice wine fermented with 1%, 3%, and 5% *P. cuspidatum* were 6.93, 107.29, 189.63, and 293.92 ppm vitamin C equivalent, respectively. The reducing power increased with the amount of *P. cuspidatum*, similar to the results of DPPH radical scavenging activity.

Another antioxidant system was investigated by determining the chelating power of the ferrous ion. Ferrozine can form complexes with the ferrous ion, resulting in a red color. In the presence of chelating agents, complex formation is interrupted, decreasing the red color. The chelation capacity of rice wine increased with the amount of *P. cuspidatum*. Rice wine fermented with *P. cuspidatum* showed higher ferrous ion chelating activity (73.05~80.91%) than control (69.36%). In general, free transition metals like iron react with either hydrogen or lipid peroxides to produce hydroxyl radical compounds and alkoxyl radicals, also known as the Fenton reaction. These radicals are extremely reactive and will significantly accelerate oxidative degradation.

In summary, the antioxidant activity (based on DPPH radical scavenging activity and reducing capacity) of the resveratrol-enriched rice wine was stronger than rice wine.

3.4. Clarification and Sterilization of Resveratrol-Enriched Rice Wine

Traditionally, boiling is used to kill microorganisms and deactivate enzymes in rice wine to protect it against rancidity and deterioration. UF is a new technology widely used by the food industry in recent years for juice clarification [54,55], but there are few reports on wine clarification. Macromolecular substances, like microorganisms, enzymes, and turbid substances, cannot pass through the UF membrane and are therefore removed from the rice wine. Aroma components and smaller nutrient molecules that can pass through the UF membrane are retained. The resveratrol-enriched rice wine was ultrafiltered with a tangential flow filtration system equipped with a hollow-fiber filter module with a molecular weight cut-off of 3 kDa or 10 kDa. The effect of the UF process on rice wine quality is shown in Table 2. UF did not affect the ethanol content, total acidity, pH, reducing sugar content, or amino acid content of the rice wine. These parameters were, respectively, about 14% ethanol, 4 g/L total acidity, 3.7 mg/mL reducing sugars, and 1.1 mg/mL amino acids. However, the UF-treated rice wine had better clarity (transmittance at 680 nm was more than 99%) than boiled rice wine (91.4%), as shown in Table 2. The transmittance shows a large difference between the UF-treated and boiled rice wine. Li et al. have shown that suspended solids and juice turbidity can be almost entirely removed by the UF membrane, but the pH, acidity, sugar, and soluble solid content of the juice showed no significant changes [56]. Furthermore, the microbiological characteristics of the UF-treated resveratrol-enriched rice wine were in agreement with food safety regulations, since the number of aerobic plate count was not detected, as shown in Table 2. These results indicate that UF could be an alternative process for clarification and sterilization. UF has been used to clarify nixtamalization waste waters for the recovery of phenolic compounds [57]. In our study, the 3 kDa UF process slightly decreased the amount of piceid and resveratrol, but the concentration of piceid and resveratrol did not change by the 10 kDa UF process. The resveratrol-enriched rice wine obtained from 10 kDa UF contains 120 mg/L piceid and 84 mg/L resveratrol, which is at least 23 times greater than the resveratrol concentration of red wine (~1.6 to 3.6 mg/L resveratrol) [17]. As such, resveratrol-enriched rice wine would have a beneficial effect on physical health.

Table 2. Quality characteristics of resveratrol-enriched rice wine treated by boiling and ultrafiltration.

Item	Boiling	Ultrafiltration	
		10 KD	**3 KD**
Ethanol (%; *v/v*)	14.22 ± 0.66	14.40 ± 0.24	14.50 ± 0.92
Piceid (mg/L)	120.10 ± 6.07	120.67 ± 3.05	104.75 ± 2.51
Resveratrol (mg/L)	82.50 ± 2.52	84.93 ± 0.47	80.58 ± 0.43
Clarity (%T)	91.4 ± 0.00	99.35 ± 0.05	99.60 ± 0.05
Aerobic plate count (CFU/mL)	N.D. [1]	N.D.	N.D.
Total acidity (g/L)	3.83 ± 0.23	4.05 ± 0.00	4.05 ± 0.00
pH	3.66	3.65	3.65
Reducing sugars (mg/mL)	3.59 ± 0.08	4.09 ± 0.19	3.55 ± 0.21
Amino acids (mg/mL)	1.11 ± 0.03	1.04 ± 0.00	1.05 ± 0.02

[1] Not detected.

3.5. Storage Study

The UF-treated rice wine was stored for 4 weeks to check its shelf life. Polymerization of phenolic compounds and interactions with other components (e.g., proteins or sugars) can lead to increased turbidity [58–60]. The rates of these reactions slow in low temperature conditions. Therefore, the UF-treated rice wine was stored at a lower temperature to keep the turbidity within a minimal range. Various parameters, including ethanol, piceid, and resveratrol levels, clarity, aerobic plate count, total acidity, pH, total reducing sugars, and total amino acids, were determined at the onset and after each consecutive week. As the results in Table 3 show, the quality of rice wine remained almost the same for 4 weeks. The quality indices (ethanol, piceid, and resveratrol) of the UF-treated rice wine showed no significant change after four weeks of storage, and therefore it can be safely asserted that the UF-treated rice wine can be stored for four weeks at normal refrigeration temperature (4 °C) without any additives or preservatives.

Table 3. Quality characteristics of resveratrol-enriched rice wine treated by boiling and ultrafiltration process after storage.

Treat	No. of Weeks	Ethanol (%)	Piceid (mg/L)	Resveratrol (mg/L)	Clarity (%T)	APC (CFU/mL)	Total Acidity (g/L)	pH	Reducing Sugars (mg/mL)	Amino Acids (mg/mL)
Boiling	1	14.8 ± 1.9 [a]	120.9 ± 0.9 [a]	84.1 ± 2.6 [a]	89.8 ± 0.1 [b]	N.D.[1]	3.38 ± 0.2 [a]	3.60	3.7 ± 0.2 [b]	1.55 ± 0.02 [d]
	2	14.8 ± 0.5 [a]	117.9 ± 8.2 [a]	85.8 ± 4.1 [a]	91.0 ± 0.0 [a]	N.D.	4.05 ± 0.0 [a]	3.69	4.1 ± 0.1 [a]	1.97 ± 0.01 [a]
	3	14.6 ± 0.0 [a]	113.2 ± 9.3 [a]	79.1 ± 1.3 [a]	88.6 ± 0.1 [c]	N.D.	4.05 ± 0.0 [a]	3.60	4.0 ± 0.1 [a]	1.75 ± 0.01 [b]
	4	14.2 ± 0.3 [a]	113.4 ± 1.0 [a]	80.0 ± 2.1 [a]	90.0 ± 0.0 [b]	N.D.	4.73 ± 0.7 [a]	3.60	4.0 ± 0.1 [ab]	1.67 ± 0.01 [c]
UF-10K	1	14.1 ± 0.7 [a]	120.4 ± 8.0 [a]	83.9 ± 0.8 [a]	99.3 ± 0.0 [a]	N.D.	3.60 ± 0.0 [c]	3.60	4.3 ± 0.2 [a]	1.48 ± 0.02 [d]
	2	14.2 ± 0.7 [a]	122.8 ± 6.5 [a]	85.2 ± 3.7 [a]	98.7 ± 0.1 [c]	N.D.	4.05 ± 0.0 [b]	3.62	4.4 ± 0.1 [a]	1.95 ± 0.02 [a]
	3	14.6 ± 0.3 [a]	120.9 ± 8.2 [a]	81.7 ± 2.4 [a]	99.0 ± 0.1 [b]	N.D.	4.05 ± 0.0 [b]	3.61	4.4 ± 0.1 [a]	1.74 ± 0.02 [b]
	4	14.2 ± 0.9 [a]	116.9 ± 2.1 [a]	82.6 ± 0.6 [a]	98.6 ± 0.0 [c]	N.D.	4.95 ± 0.0 [a]	3.59	4.1 ± 0.2 [a]	1.66 ± 0.03 [c]
UF-3K	1	14.1 ± 2.3 [a]	107.9 ± 4.2 [a]	80.9 ± 1.1 [a]	99.6 ± 0.1 [a]	N.D.	3.60 ± 0.0 [b]	3.61	3.9 ± 0.1 [a]	1.47 ± 0.02 [d]
	2	14.2 ± 1.6 [a]	101.0 ± 5.6 [a]	78.0 ± 0.4 [a]	98.5 ± 0.1 [b]	N.D.	4.05 ± 0.0 [a]	3.63	4.0 ± 0.1 [a]	1.94 ± 0.00 [a]
	3	14.8 ± 1.7 [a]	109.8 ± 12 [a]	73.9 ± 0.2 [a]	98.7 ± 0.1 [b]	N.D.	4.05 ± 0.0 [a]	3.61	4.1 ± 0.1 [a]	1.74 ± 0.02 [b]
	4	14.1 ± 0.6 [a]	97.2 ± 0.0 [a]	75.1 ± 3.4 [a]	97.8 ± 0.1 [c]	N.D.	4.28 ± 0.2 [a]	3.57	4.0 ± 0.2 [a]	1.62 ± 0.04 [c]

[1] Not detected. [a,b,c,d] When the quality characteristics of rice wine with the same treatment have significant differences at p-value less than 0.05, homogeneous groups in each variable are identified by the same superscript letter according to the LSD test. APC: Aerobic plate count.

4. Conclusions

This study established a new process to combine rice wine fermentation, the extraction of phenolic compounds, and the biotransformation of piceid to resveratrol to obtain functional rice wine. This novel resveratrol-enriched rice wine has enhanced antioxidant activities, including increased free radical-scavenging activity, ferric ion reducing power, and ferrous ion chelating activity. The results confirm that UF has great potential for the clarification and sterilization of rice wine as it retains bioactive compounds. This is because UF operates at room temperature, no chemicals are used, no phase changes occur, and it has a high recovery yield. The UF-treated clarified product presents good quality characteristics, and also meets microbiological safety requirements. Therefore, this novel product could be consumed as a resveratrol-rich functional wine or a natural herb-fermented rice wine.

Author Contributions: Investigation, K.-R.Y. and H.-C.Y.; Resources, C.-Y.H., J.-M.K., and C.C.; Supervision, C.-H.K.; Validation, C.-H.K.; Writing—original draft, C.-H.K.; Writing—review and editing, C.-J.S.

Funding: This research was funded by the Ministry of Science and Technology of Taiwan, grant number MOST 107-2221-E-992-031.

Conflicts of Interest: The authors declare no conflict of interest.

References

1. Hasan, M.; Bae, H. An overview of stress-induced resveratrol synthesis in grapes: Perspectives for resveratrol-enriched grape products. *Molecules* **2017**, *22*, 294. [CrossRef] [PubMed]
2. Neto, C.C. Cranberry and blueberry: Evidence for protective effects against cancer and vascular diseases. *Mol. Nutr. Food Res.* **2007**, *51*, 652–664. [CrossRef] [PubMed]
3. Chen, R.-S.; Wu, P.-L.; Chiou, R.Y.-Y. Peanut roots as a source of resveratrol. *J. Agric. Food Chem.* **2002**, *50*, 1665–1667. [CrossRef] [PubMed]
4. Sun, Y.; Zhou, Q.-M.; Lu, Y.-Y.; Zhang, H.; Chen, Q.-L.; Zhao, M.; Su, S.-B. Resveratrol inhibits the migration and metastasis of MDA-MB-231 human breast cancer by reversing TGF-β1-induced epithelial-mesenchymal transition. *Molecules* **2019**, *24*, 1131. [CrossRef] [PubMed]
5. Lee, C.-C.; Chen, Y.-T.; Chiu, C.-C.; Liao, W.-T.; Liu, Y.-C.; Wang, H.-M.D. *Polygonum cuspidatum* extracts as bioactive antioxidaion, anti-tyrosinase, immune stimulation and anticancer agents. *J. Biosci. Bioeng.* **2015**, *119*, 464–469. [CrossRef] [PubMed]
6. Rubio-Ruiz, M.E.; Guarner-Lans, V.; Cano-Martínez, A.; Díaz-Díaz, E.; Manzano-Pech, L.; Gamas-Magaña, A.; Castrejón-Tellez, V.; Tapia-Cortina, C.; Pérez-Torres, I. Resveratrol and quercetin administration improves antioxidant defenses and reduces fatty liver in metabolic syndrome rats. *Molecules* **2019**, *24*, 1297. [CrossRef] [PubMed]
7. Rodríguez-Pérez, C.; Segura-Carretero, A.; del Mar Contreras, M. Phenolic compounds as natural and multifunctional anti-obesity agents: A review. *Crit. Rev. Food Sci Nutr.* **2019**, *59*, 1212–1229.
8. Chimento, A.; De Amicis, F.; Sirianni, R.; Sinicropi, M.S.; Puoci, F.; Casaburi, I.; Saturnino, C.; Pezzi, V. Progress to improve oral bioavailability and beneficial effects of resveratrol. *Int. J. Mol. Sci.* **2019**, *20*, 1381. [CrossRef]
9. Hasan, M.M.; Yun, H.-K.; Kwak, E.-J.; Baek, K.-H. Preparation of resveratrol-enriched grape juice from ultrasonication treated grape fruits. *Ultrason. Sonochem.* **2014**, *21*, 729–734. [CrossRef]
10. Wada, M.; Kido, H.; Ohyama, K.; Ichibangase, T.; Kishikawa, N.; Ohba, Y.; Nakashima, M.; Kuroda, N.; Nakashima, K. Chemiluminescent screening of quenching effects of natural colorants against reactive oxygen species: Evaluation of grape seed, monascus, gardenia and red radish extracts as multi-functional food additives. *Food Chem.* **2007**, *101*, 980–986. [CrossRef]
11. Katalinić, V.; Možina, S.S.; Skroza, D.; Generalić, I.; Abramovič, H.; Miloš, M.; Ljubenkov, I.; Piskernik, S.; Pezo, I.; Terpinc, P. Polyphenolic profile, antioxidant properties and antimicrobial activity of grape skin extracts of 14 *Vitis vinifera* varieties grown in Dalmatia (Croatia). *Food Chem.* **2010**, *119*, 715–723. [CrossRef]
12. Bavaresco, L.; Lucini, L.; Busconi, M.; Flamini, R.; De Rosso, M. Wine resveratrol: From the ground up. *Nutrients* **2016**, *8*, 222. [CrossRef] [PubMed]

13. Arranz, S.; Chiva-Blanch, G.; Valderas-Martínez, P.; Medina-Remón, A.; Lamuela-Raventós, R.M.; Estruch, R. Wine, beer, alcohol and polyphenols on cardiovascular disease and cancer. *Nutrients* **2012**, *4*, 759–781. [CrossRef] [PubMed]

14. Ragusa, A.; Centonze, C.; Grasso, M.E.; Latronico, M.F.; Mastrangelo, P.F.; Sparascio, F.; Maffia, M. HPLC analysis of phenols in negroamaro and primitivo red wines from Salento. *Foods* **2019**, *8*, 45. [CrossRef] [PubMed]

15. Kallithraka, S.; Arvanitoyannis, I.; El-Zajouli, A.; Kefalas, P. The application of an improved method for trans-resveratrol to determine the origin of Greek red wines. *Food Chem.* **2001**, *75*, 355–363. [CrossRef]

16. Ragusa, A.; Centonze, C.; Grasso, M.; Latronico, M.; Mastrangelo, P.; Sparascio, F.; Fanizzi, F.; Maffia, M. A comparative study of phenols in Apulian Italian Wines. *Foods* **2017**, *6*, 24. [CrossRef]

17. Tosun, I.; Inkaya, A.N. Resveratrol as a health and disease benefit agent. *Food Rev. Int.* **2009**, *26*, 85–101. [CrossRef]

18. Zhang, H.; Li, C.; Kwok, S.-T.; Zhang, Q.-W.; Chan, S.-W. A review of the pharmacological effects of the dried root of *Polygonum cuspidatum* (Hu Zhang) and its constituents. *Evid. Based Complementary Altern. Med.* **2013**, *2013*, 208349. [CrossRef]

19. Chen, B.-Y.; Kuo, C.-H.; Liu, Y.-C.; Ye, L.-Y.; Chen, J.-H.; Shieh, C.-J. Ultrasonic-assisted extraction of the botanical dietary supplement resveratrol and other constituents of *Polygonum cuspidatum. J. Nat. Prod.* **2012**, *75*, 1810–1813. [CrossRef]

20. Liu, Z.; Wei, F.; Chen, L.-J.; Xiong, H.-R.; Liu, Y.-Y.; Luo, F.; Hou, W.; Xiao, H.; Yang, Z.-Q. In vitro and in vivo studies of the inhibitory effects of emodin isolated from *Polygonum cuspidatum* on Coxsakievirus B4. *Molecules* **2013**, *18*, 11842–11858. [CrossRef]

21. Xiang, M.-X.; Xu, Z.; Su, H.-W.; Hu, J.; Yan, Y.-J. Emodin-8-O-β-D-glucoside from *Polygonum Amplexicaule* D. Don var. Sinense Forb. promotes proliferation and differentiation of osteoblastic MC3T3-E1 Cells. *Molecules* **2011**, *16*, 728–737. [CrossRef] [PubMed]

22. Li, H.-L.; Chen, H.-L.; Li, H.; Zhang, K.-L.; Chen, X.-Y.; Wang, X.-W.; Kong, Q.-Y.; Liu, J. Regulatory effects of emodin on NF-κB activation and inflammatory cytokine expression in RAW 264.7 macrophages. *Int. J. Mol. Med.* **2005**, *16*, 41–47. [CrossRef] [PubMed]

23. Froldi, G.; Baronchelli, F.; Marin, E.; Grison, M. Antiglycation activity and HT-29 cellular uptake of aloe-Emodin, aloin, and aloe arborescens leaf extracts. *Molecules* **2019**, *24*, 2128. [CrossRef] [PubMed]

24. Kuo, C.-H.; Chen, B.-Y.; Liu, Y.-C.; Chang, C.-M.; Deng, T.-S.; Chen, J.-H.; Shieh, C.-J. Optimized ultrasound-assisted extraction of phenolic compounds from *Polygonum cuspidatum. Molecules* **2014**, *19*, 67–77. [CrossRef] [PubMed]

25. Lin, J.-A.; Kuo, C.-H.; Chen, B.-Y.; Li, Y.; Liu, Y.-C.; Chen, J.-H.; Shieh, C.-J. A novel enzyme-assisted ultrasonic approach for highly efficient extraction of resveratrol from *Polygonum cuspidatum. Ultrason. Sonochem.* **2016**, *32*, 258–264. [CrossRef] [PubMed]

26. Zhang, C.; Li, D.; Yu, H.; Zhang, B.; Jin, F. Purification and characterization of piceid-β-D-glucosidase from *Aspergillus oryzae. Process Biochem.* **2007**, *42*, 83–88. [CrossRef]

27. Kuo, C.-H.; Chen, B.-Y.; Liu, Y.-C.; Chen, J.-H.; Shieh, C.-J. Production of resveratrol by piceid deglycosylation using cellulase. *Catalysts* **2016**, *6*, 32. [CrossRef]

28. Park, H.-J.; Lee, S.; Song, S.; Kim, Y.-S. Characterization of volatile components in Makgeolli, a traditional Korean rice wine, with or without pasteurization, during storage. *Molecules* **2013**, *18*, 5317–5325. [CrossRef]

29. Mo, X.; Xu, Y.; Fan, W. Characterization of aroma compounds in Chinese rice wine Qu by solvent-assisted flavor evaporation and headspace solid-phase microextraction. *J. Agric. Food Chem.* **2010**, *58*, 2462–2469. [CrossRef]

30. Xie, G.F.; Li, W.J.; Lu, J.; Cao, Y.; Fang, H.; Zou, H.J.; Hu, Z.M. Isolation and identification of representative fungi from Shaoxing rice wine wheat Qu using a polyphasic approach of culture-based and molecular-based methods. *J. Inst. Brew* **2007**, *113*, 272–279. [CrossRef]

31. Kuo, C.-H.; Shieh, C.-J.; Huang, S.-M.; Wang, H.-M.D.; Huang, C.-Y. The effect of extrusion puffing on the physicochemical properties of brown rice used for saccharification and Chinese rice wine fermentation. *Food Hydrocolloids* **2019**, *94*, 363–370. [CrossRef]

32. El-Deen, A.M.N.; Shata, H.M.A.H.; Farid, M.A.F. Improvement of β-glucosidase production by co-culture of *Aspergillus niger* and *A. oryzae* under solid state fermentation through feeding process. *Ann. Microbiol.* **2014**, *64*, 627–637. [CrossRef]

33. Karmakar, M.; Ray, R.R. A statistical approach for optimization of simultaneous production of β-glucosidase and endoglucanase by *Rhizopus oryzae* from solid-state fermentation of water hyacinth using central composite design. *Biotechnol. Res. Int.* **2011**, *2011*, 574983. [CrossRef] [PubMed]

34. Tazawako Beer. Available online: https://www.warabi.or.jp/beer/dragon.html (accessed on 24 June 2019).

35. Kuo, C.H.; Chen, C.C.; Chiang, B.H. Process characteristics of hydrolysis of chitosan in a continuous enzymatic membrane reactor. *J. Food Sci.* **2004**, *69*, 332–337. [CrossRef]

36. Wei, J.T.; Chiang, B.H. Bioactive peptide production by hydrolysis of porcine blood proteins in a continuous enzymatic membrane reactor. *J. Sci. Food Agric.* **2009**, *89*, 372–378. [CrossRef]

37. Conidi, C.; Destani, F.; Cassano, A. Performance of hollow fiber ultrafiltration membranes in the clarification of blood orange juice. *Beverages* **2015**, *1*, 341–353. [CrossRef]

38. Jiao, A.; Xu, X.; Jin, Z. Research progress on the brewing techniques of new-type rice wine. *Food Chem.* **2017**, *215*, 508–515. [CrossRef] [PubMed]

39. Bhattacharjee, C.; Saxena, V.K.; Dutta, S. Analysis of fouling and juice quality in crossflow ultrafiltration of watermelon juice. *Food Sci. Technol.* **2018**, *38*, 71–76. [CrossRef]

40. Debien, I.C.D.N.; Gomes, M.T.D.M.S.; Ongaratto, R.S.; Viotto, L.A. Ultrafiltration performance of PVDF, PES, and cellulose membranes for the treatment of coconut water (*Cocos nucifera* L.). *Food Sci. Technol.* **2013**, *33*, 676–684. [CrossRef]

41. Huang, C.-Y.; Kuo, C.-H.; Chen, P.-W. Compressional-puffing pretreatment enhances neuroprotective effects of fucoidans from the brown seaweed Sargassum hemiphyllum on 6-hydroxydopamine-induced apoptosis in SH-SY5Y cells. *Molecules* **2018**, *23*, 78. [CrossRef]

42. Conde, E.; Cara, C.; Moure, A.; Ruiz, E.; Castro, E.; Domínguez, H. Antioxidant activity of the phenolic compounds released by hydrothermal treatments of olive tree pruning. *Food Chem.* **2009**, *114*, 806–812. [CrossRef]

43. Wu, H.-C.; Pan, B.S.; Chang, C.-L.; Shiau, C.-Y. Low-molecular-weight peptides as related to antioxidant properties of chicken essence. *J. Food Drug Anal.* **2005**, *13*, 176–183.

44. Kuo, C.-H.; Lin, P.-J.; Wu, Y.-Q.; Ye, L.-Y.; Yang, D.-J.; Shieh, C.-J.; Lee, C.-K. Simultaneous saccharification and fermentation of waste textiles for ethanol production. *BioResources* **2014**, *9*, 2866–2875. [CrossRef]

45. Church, F.C.; Swaisgood, H.E.; Porter, D.H.; Catignani, G.L. Spectrophotometric assay using o-phthaldialdehyde for determination of proteolysis in milk and isolated milk proteins. *J. Dairy Sci.* **1983**, *66*, 1219–1227. [CrossRef]

46. Wang, H.; Liu, L.; Guo, Y.-X.; Dong, Y.-S.; Zhang, D.-J.; Xiu, Z.-L. Biotransformation of piceid in *Polygonum cuspidatum* to resveratrol by *Aspergillus oryzae*. *Appl. Microbiol. Biotechnol.* **2007**, *75*, 763. [CrossRef] [PubMed]

47. Orozco, H.; Matallana, E.; Aranda, A. Two-carbon metabolites, polyphenols and vitamins influence yeast chronological life span in winemaking conditions. *Microb. Cell Fact.* **2012**, *11*, 104. [CrossRef] [PubMed]

48. Ahn, H.; You, H.; Park, M.; Johnston, T.; Ku, S.; Ji, G. Biocatalysis of platycoside E and platycodin D3 using fungal extracellular B-glucosidase responsible for rapid platycodin D production. *Int. J. Mol. Sci.* **2018**, *19*, 2671. [CrossRef]

49. La Torre, G.L.; Laganà, G.; Bellocco, E.; Vilasi, F.; Salvo, F.; Dugo, G. Improvement on enzymatic hydrolysis of resveratrol glucosides in wine. *Food Chem.* **2004**, *85*, 259–266. [CrossRef]

50. Todaro, A.; Palmeri, R.; Barbagallo, R.N.; Pifferi, P.G.; Spagna, G. Increase of trans-resveratrol in typical Sicilian wine using β-glucosidase from various sources. *Food Chem.* **2008**, *107*, 1570–1575. [CrossRef]

51. Huang, C.-Y.; Kuo, C.-H.; Lee, C.-H. Antibacterial and antioxidant capacities and attenuation of lipid accumulation in 3T3-L1 adipocytes by low-molecular-weight fucoidans prepared from compressional-puffing-pretreated sargassum crassifolium. *Mar. Drugs* **2018**, *16*, 24.

52. Sae-Lee, N.; Kerdchoechuen, O.; Laohakunjit, N. Enhancement of phenolics, resveratrol and antioxidant activity by nitrogen enrichment in cell suspension culture of *Vitis vinifera*. *Molecules* **2014**, *19*, 7901–7912. [CrossRef] [PubMed]

53. Pan, Y.; Zhang, X.; Wang, H.; Liang, Y.; Zhu, J.; Li, H.; Zhang, Z.; Wu, Q. Antioxidant potential of ethanolic extract of *Polygonum cuspidatum* and application in peanut oil. *Food Chem.* **2007**, *105*, 1518–1524. [CrossRef]

54. Loizzo, M.R.; Sicari, V.; Tundis, R.; Leporini, M.; Falco, T.; Calabrò, V. The influence of ultrafiltration of *Citrus limon* L. Burm. cv Femminello comune juice on its chemical composition and antioxidant and hypoglycemic properties. *Antioxidants* **2019**, *8*, 23. [CrossRef] [PubMed]

55. Abadi, S.; Mahdi, M.; Kashani Nejad, M.; Mahunak, S.; Reza, A. Effect of ultrafiltration process on quality characteristics of sour orange juice. *Food Sci. Technol.* **2016**, *13*, 119–129.
56. Li, J.; Wang, Z.; Ge, Y.; Sun, Q.; Hu, X. Clarification and sterilization of raw depectinized apple juice by ceramic ultrafiltration membranes. *J. Sci. Food Agric.* **2006**, *86*, 148–155. [CrossRef]
57. Castro-Muñoz, R.; Barragán-Huerta, B.E.; Yáñez-Fernández, J. The use of nixtamalization waste waters clarified by ultrafiltration for production of a fraction rich in phenolic compounds. *Waste Biomass Valorization* **2016**, *7*, 1167–1176.
58. Štajner, D.; Milošević, M.; Popović, B. Irradiation effects on phenolic content, lipid and protein oxidation and scavenger ability of soybean seeds. *Int. J. Mol. Sci.* **2007**, *8*, 618–627. [CrossRef]
59. McRae, J.M.; Kennedy, J.A. Wine and grape tannin interactions with salivary proteins and their impact on astringency: A review of current research. *Molecules* **2011**, *16*, 2348–2364. [CrossRef]
60. Lorrain, B.; Ky, I.; Pechamat, L.; Teissedre, P.-L. Evolution of analysis of polyhenols from grapes, wines, and extracts. *Molecules* **2013**, *18*, 1076–1100. [CrossRef]

MDPI

St. Alban-Anlage 66

4052 Basel

Switzerland

Tel. +41 61 683 77 34

Fax +41 61 302 89 18

www.mdpi.com

Foods Editorial Office

E-mail: foods@mdpi.com

www.mdpi.com/journal/foods